LNG

LNG

A Level-Headed Look
at the
Liquefied Natural Gas
Controversy

Virginia L. Thorndike

DOWN
EAST
BOOKS

Printed at Versa Press, E. Peoria, Ill.

2   4   5   3   1

LIBRARY OF CONGRESS CATALOGING-IN-PUBLICATION DATA

Thorndike, Virginia L.
    LNG : a level-headed look at the liquefied natural gas
controversy / by Virginia L. Thorndike.
        p.   cm.
    Includes bibliographical references and index.
    ISBN-13: 978-0-89272-701-8 (trade pbk. : alk. paper)
    ISBN-10: 0-89272-701-2
    1. Liquefied natural gas. I. Title.
    TP359.L5T46 2007
    333.8'233--dc22
                                                2006034552

DOWN EAST BOOKS / A division of Down East Enterprise, Inc.,
Publisher of Down East, the Magazine of Maine

www.downeastbooks.com / Book orders: 1-800-685-7962
Distributed to the trade by National Book Network

To all of you who really care
about doing things right,
and one of those people in particular

Contents

COURTESY OF CAPTAIN DICK JOHNSTON

Taking a Dispassionate Look at LNG

EVERY time I told people involved in any way with liquefied natural gas (LNG) that my goal was to write a dispassionate view of the issues surrounding LNG, they laughed, often with outright guffaw. Within the world of LNG, there is no dispassion. All participants know what they think, and they're not likely to be influenced by another view. But, what about the rest of us? We hear talk of LNG as if it would be the saving of the world, and we also hear heartfelt diatribes from others who see LNG as being the cause of, if not the end of humanity, at least of their local way of life.

"We're not a very good target for terrorists," said the prospective developer of one project, tapping into a very important issue in many people's minds. Much print and public discussion is given to the question of whether or not an LNG tanker or plant would be attractive to malefactors with a desire to make a big statement. Unfortunately, in this particular case, the developer had already misled his listeners when he reportedly explained that in the event of an attempted explosion, the gas would evaporate and emit carbon monoxide and methane into the air.[1] While he is right that natural gas can indeed evaporate away (though one wonders where the carbon monoxide came from), there's more to the story. His disingenuousness does his industry no good. It is important that people present the facts honestly.

It is my intent to take a look at the world of LNG in a straightforward and matter-of-fact manner, to show what it is and how it works, in the hope of defusing some of the frenzy and hype surrounding proposals for new import facilities. My entrée into the subject was through tugboats—tugboats play a key role in the importation of LNG, and their operators tend to be supportive of the industry, both for financial reasons and because of their belief in their own abilities. Their positive views counterbalanced any anti-LNG tendency I might have started with. I now, truly, have no axes to grind. I have listened to many of the most serious and knowledgeable people involved in every aspect of the field, and although I am no scientist and do not even remotely resemble an engineer, I believe I have made some sense of the issues in front of us today.

In North America, we use tremendous and ever-increasing amounts of energy, and no sign of that trend changing is apparent, no matter how obvious the need for change may seem. Significant policy changes will need to be made on the national level, either to reduce our use of energy or to come up with a more permanent technological solution, or, most likely, both. While that policy choice is undoubtedly the most important issue for consideration, it is beyond the scope of this book.

At this moment, halfway through the first decade of the twenty-first century, more than twenty new LNG projects are under consideration in the United States, and, in fact, another eighteen already have been approved by FERC (the Federal Energy Regulatory Commission)—or the Coast Guard (USCG), in the case of offshore facilities. How many of these will get built is an entirely different issue, one that will ultimately be decided by the financial people. The future and current-day economics (and politics) of energy in general and of LNG in particular are also well beyond my capabilities. In addition, I make little attempt to analyze technical (and political) issues such as whether there is a sufficient supply of natural gas available to justify construction of LNG facilities in the United States or elsewhere. I must assume that the companies who wish to build such things know more about this than I do. They certainly have a much larger stake in making the right assessment.

It does seem, however, that anyone who's looking to LNG to lower the price of natural gas significantly is likely to be disappointed. LNG has too many built-in expenses—liquefaction, transportation, and regasifying. While in the past there have been some very good deals on natural gas overseas, which allowed the other costs of LNG to be absorbed, the nations and companies that have the resource are no longer as enthusiastic about selling it so cheaply. Who can blame them?

LNG cannot be the answer to all our problems, no matter what, but it is being offered as one of several answers for at least a few more decades. I make no recommendations about particular new projects one way or the other. When I began researching the controversy surrounding LNG, I wanted to take a look at the facts that will help people figure out how they should feel about those projects both in general and in particular. But in doing that, I came up with some conclusions of my own. These views will be apparent.

U.S. Energy Basics[2]

SINCE forever, we humans have looked for ways to increase our physical capabilities. We started by using levers and wheels and beasts of burden—and, of course, fire came along at some point, with wood as fuel—and later, we coerced wind and water to work for us as well. We have only gotten more clever over time, and found more and more ways to do more and more, using less and less of our own muscle power. Never so quickly had this trend increased as during the nineteenth and early twentieth centuries, when new energy sources were discovered and made manageable. In that time, our lives changed entirely. In the United States in particular, during the course of just a hundred years, fossil fuels—hydrocarbons—and their accompanying industrial revolution transformed a rural society of largely self-sufficient farmers with no international significance into an urban, international superpower. And now, dependent as we are on other nations for much of our energy supply, we face the reversing of at least some aspects of that progression.

In this country during 1900, an average of 122 million Btus of energy was expended for each one of us; by 1949, the figure was 215 million Btus, and by 1973, that per-capita usage had increased to 350 million.[3] (A British thermal unit (Btu) is the amount of heat required to raise the temperature of one pound of water by one degree Fahrenheit. This term is commonly used in the United States if nowhere else. For comparison, 96 pounds of coal are required to create 1 million Btus, or 7.21 gallons of diesel or heating oil, or 8.07 gallons of gasoline, or 970 cubic feet of natural gas, or 293 kilowatt-hours of electricity. Municipal solid waste and wood or wood waste average a million Btus per 200 pounds, and a million Btus from discarded vehicle tires would require only 74 pounds.)

Aside from actual drops in times of energy "crisis," the rate of use per person has stayed fairly level since the seventies. But between 1949 and 2000, the United States came close to doubling its population, from 149 million to 281 million, increasing the call for energy simply by adding to the number of people demanding it.

For many years, this country's primary source of energy was wood; by the middle of the nineteenth century, this was changing to coal. Coal was plentiful, and, until miners got the idea that they shouldn't work for pittances underground in extremely dangerous conditions, it was inexpensive. But after World War II, petroleum began to provide a larger and larger share of our energy supply, and in 1951, for the first time, petroleum became dominant in the United States, which was then the world leader in oil production. Coal, increasingly and ever more efficiently mined from the surface, particularly west of the Mississippi, was still in demand for electrical generation if nothing else, and during the oil crisis of the 1970s, coal again took over the top spot in U.S. energy.

Until the late 1950s, as a nation, we were still self-sufficient for our energy requirements, but since that time, we've imported more and more, mostly in the form of oil. Since its peak in 1970, our own oil production has been dropping steadily. We imported more oil than we produced for the first time in 1994, and by 2000 we were bringing in more than twice the oil we had imported just fifteen years earlier, some eleven million barrels a day. Just over half of that comes from OPEC nations Saudi Arabia, Venezuela, and Nigeria. And after a relatively flat crude oil cost that started in the mid-seventies and lasted many decades, more recently the price in constant dollars has bounced all over the place. Since 1994, the yearly average for crude oil has ranged from a low of $15.99 a barrel[4] to over $75 in the summer of 2006. Transportation, of course, accounts for a huge portion of our energy use, nearly all of it petroleum in the form of diesel or gasoline. Currently, transportation accounts for about twice the petroleum consumption of residential, commercial, and industrial purposes combined.

It's true, as some say, that, despite our consumption, the earth will never run out of oil. In fact, a number of earlier predictions of a time of diminishing production have come and gone, and still more petroleum is pumped every year. However, thinking that there will always be a ready supply of oil is rash. Even with increasing prices, a time will come when the oil remaining worldwide will not be economically feasible to extract. Sooner or later, its production, like that of all the nonrenewable resources, will maximize and then fall off until it disappears altogether. We read more and more about this concept, called "Peak Oil." This theory predicts a maximizing of oil production worldwide in the near future, to be followed by a rapid decline.

In 2000, the U.S. Geological Survey (USGS) released what the U.S. Department of Energy (DOE) calls "the most thorough and methodologically modern assessment of world crude oil and natural gas resources ever attempted." Forty geoscientists, many with industry backgrounds, conducted the five-year study, which was reviewed at every stage by others involved in the business. The Energy Information Administration (EIA) then produced an analysis of the resources.

The EIA and USGS studies estimated the peak of oil production at some time between 2026 and 2047, depending on annual demand and, therefore, production growth rate, with essentially no further production after 2100 or 2125 at the latest. The EIA's study concludes: "The world production peak for conventionally reservoired crude is unlikely to be 'right around the corner' as so many other estimators have been predicting. Our analysis shows that it will be closer to the middle of the 21st century than to its beginning. Given the long lead times required for significant mass-market penetration of new energy technologies, this result in no way justifies complacency about both supply-side and demand-side research and development."[5]

If they're right, we have about a hundred years until it's all done, and maybe only twenty or forty until peak production is reached. It might be sooner. In 2004, China suddenly used a tremendous amount more oil than it ever had before; India and China are obviously industrializing at a very rapid pace. One can question the EIA's definition of "right around the corner." Even twenty to fifty years isn't such a long time.

As for natural gas, for several years the United States has produced about eighteen trillion cubic feet (Tcf) of natural gas a year, and we have used about twenty-two Tcf annually. By far the greatest share of our imported natural gas comes from Canada (thereby giving Canada the honor of being second only to Russia as the largest exporter of that commodity).

The future of natural gas is subject to just as much speculation as is that of petroleum. As of the beginning of 2004, in conjunction with *Oil & Gas Journal*, the EIA estimated the proved world reserves to be around 6,079 Tcf, about three-quarters of which is in the Middle East, Eastern Europe, and the former USSR. (Proved reserves are those resources that are reasonably certain to be recoverable, given current technology and economic conditions.) Significant other resources may, in fact, be developed—new finds or, more likely, known resources that are not now economically harvestable, but may become so in the future through new technology or because of higher prices. However, there's no way to confidently make estimates of those.

Although thought to be but 4 percent of the world total, significant proved natural gas reserves still exist in North America. About 40 percent of our present resource is now "off limits," either in regions with no access, like the Rockies, or in areas that we have specifically chosen not to harvest—Alaska, the Florida side of the Gulf of Mexico, and off and along both the Atlantic and Pacific coasts. These resources are not generally included in the inventory.[6]

Imports of natural gas have stayed more balanced than have those of oil. Although we presently account for nearly a quarter of the worldwide natural gas consumption, for the last several years natural gas has made up somewhat less than a quarter of the total energy used in the United States. Manufacturing accounts for the largest share; it is utilized in plastics, fertilizer, laundry detergents, synthetic rubber, and other products. Natural gas heats over half our homes and is used in commercial buildings, too. A small percentage is used for transportation, and the amount of natural gas used to fire electrical generation plants is now about equal to that used residentially.

Even the most conservative estimates would allow for over forty-five years' supply from domestic sources at the current rate of use. However, all predictions suggest a significant increase in demand in the United States in the next quarter century, particularly for electrical generation, because recently built electrical plants throughout the country are primarily gas fired. (Former Federal Regulatory Commissioner Nora Brownell points out an interesting aspect of the situation in the Northeast in particular: significant new gas-fired electric generation capability was constructed in recent years without a clear plan as to where the natural gas was coming from. Now, there's a shortage.)

The EIA predicts that in twenty years, overall demand will be 40 percent higher than today's, doubling the shortfall to about eight trillion cubic feet in North America by 2025. The bulk of that increase will have to be met from overseas as liquefied natural gas, LNG. Consequently, by 2025, almost three-quarters of the net U.S. imports of natural gas are expected to come from LNG. At the same time, worldwide demands will also increase; EIA's prediction is that planetary natural gas usage in 2025 will be nearly 70 percent higher than it was in 2002.

As for electrical generation in the United States today, very nearly half our electricity is still being produced with coal power, and nuclear plants account for almost a fifth. Renewable energy sources make up about 10 percent, between hydroelectric and others (primarily wood and waste, and to a small degree geothermal, solar, and wind). Another fifth comes from oil and natural gas, mostly the latter.

Coal's share may drop; utility companies are becoming more concerned about emissions, and coal is the worst offender in that regard. Nevertheless, coal is plentiful. At today's level of use, the 275 billion short tons of whose existence the EIA is conservatively confident would last more than two hundred years. If the trend to burn less coal continues, and more coal is identified or found economically viable, there's no saying when it would run out. If coal's contributions to air pollution aren't limited, though, perhaps no one will be around to care.

No method of producing energy is without ill effect of some kind—displacing natural forest habitat or threatening wildlife or creating heat or spent nuclear fuel that has then to be disposed of—but currently of most concern are the so-called greenhouse gases that, by blocking infrared radiation from Earth into space, are raising the temperature on the earth. (As long ago as 1272, England's King Edward I attempted to ban the burning of coal, because he recognized that in addition to providing warmth, it produced smoke that made eyes red and uncomfortable. Despite a penalty of death, the effort didn't work; wood had become too scarce and expensive. Other attempts were also made to ban or limit coal burning, but reform wasn't enacted until after a four-day smog in 1952 killed about four thousand Londoners.)[7]

Of the fossil fuels used to generate electricity, natural gas has the fewest ill effects. It burns more cleanly than coal or oil and emits far lower levels of harmful by-products into the air. Burning an equivalent energy value of either coal or oil dumps considerably more carbon dioxide into the atmosphere and vastly more nitrogen oxides, sulfur dioxide, and particulates than does the use of natural gas. No mercury is emitted from natural gas combustion, though burning natural gas does produce somewhat more carbon monoxide than burning oil does.

The only fuel powering a significant share of the country's electrical generation that has less negative impact on the atmosphere than natural gas is nuclear power, but that has its own problems as well—concerns about accidents, disposal of spent fuel rods and other radioactive waste, and the general "not in my backyard" syndrome. Still, there is increasing interest in nuclear power in the United States, even on the part of former opponents, who have recognized its potential for ecologically clean power production. But unless licensing becomes easier again, nuclear power will be slow to develop here.

In many other countries, a very significant and increasing part of their energy portfolio is in nuclear plants. As of 2005, Japan made 29 percent of its electricity from nuclear power and thirteen new plants were under construction or contract. Seventy-eight percent of France's electricity was from nuclear plants. Sweden's was over half from nuclear sources—but they're now phasing out their entire nuclear capacity because they feel it's unsafe. Yet in terms just of the numbers of kilowatt-hours generated, the United States had everyone else beat by a wide margin; in 2004, we generated more than 788 billion kWh by nuclear means, out of a world-wide total of 2,618 billion.[8]

Incidentally, electricity is a remarkably inefficient use of energy. Two of every three units of power are lost in the conversion process—that means that for every million Btus of electricity you produce, you've used three million in fuel. The rest is lost as waste heat. Five percent of the electricity finally produced is used in the generation plants, and another 9 percent is lost in transmission. Relatively little actually makes it to the end user.

The Future—Can We Go on as We Are?

Many prognostications have been made about energy, some entirely un-concerned about shortages and others foreseeing serious crunches in the immediate future. The realistic ones acknowledge, at least begrudgingly, that the United States is unlikely to change its ways any time soon. There will be more and more demand for energy in general, and that energy has to come from somewhere. There are concerns not only about quanti-ties of energy but about increasing costs and price spikes (such as after the 2005 hurricanes). Leaving politics out of it, which of course we can't, it's clear that to rely too heavily on any one source is unwise, even though some have more potential than others.

Renewables and Alternative Fuels

Renewable and alternative fuels, either nontraditional or extremely trad-itional, like wood or peat, are getting more attention again. They cer-tainly would seem to have an obvious appeal in this time when we're concerned about running out of hydrocarbons, although not all are actually increasing regularly in their contribution to the energy needs of the country. In 2003, 6.1 percent of the nation's energy came from renewables: nearly half each of biomass and hydroelectric, with a smattering of solar, geothermal, and wind power thrown in, contributing less than 0.5 percent of the total. But the actual consumption wobbles. During the fifteen years between 1989 and 2003, inexplicably it was in 1996 that renewables had the highest total contribution in Btus (7.129 quadrillion Btus), and in 2001 they had the lowest (5.272 quadrillion Btus).

Hydroelectric Power

Hydroelectric plants have a long history, but few new ones of any size are likely to be constructed in the United States, because the rivers on which they are constructed suffer severe environmental impacts. It seems nearly impossible that big dams that flood people's property the way the Tennessee Valley Authority did during the Depression would be either feasible or permitted; they are extremely expensive as well as devastating to the local environment, and public opposition is far too great. Even facilities that simply divert water from its normal course and run it through a turbine may be causing harm to fish and other aquatic species. Many of these smaller dams are being removed from rivers across the country, but as other energy sources are becoming more expensive, some hydroelectric plants might be refurbished or constructed anew. The Federal Energy Regulatory Commission (FERC) has before it a few applications for new hydro projects at existing dams, and at this writing, schemes are under consideration for damless, tide-generated facilities in down east Maine and the Canadian Maritimes. (President Franklin Roosevelt may have had a good idea when he directed the Army Corps of Engineers to build a Passamaquoddy Bay tidal energy project in the 1930s, though it didn't bear fruit during his era.)

Hydroelectric power does have the potential for some expansion, but certainly not on a scale to decrease the call for hydrocarbons.

Biomass

At this time in the United States, biomass is the other significant source of renewable energy. Wood, in many forms, municipal and other waste, and alcohol fuels, mostly ethanol, all fall into this category, along with agricultural by-products, old tires, and a few less common materials.

Ethanol, derived from corn, is being added to gasoline, and its use doubled from 1999 to 2003; by 2004, it had risen to a whopping 2 percent of the oil used in the transportation sector. Even doubling or tripling this contribution would make little dent in the total picture, but it is a goal worth working toward and efforts are being made for this to happen. Estimates of the energy balance of corn-based ethanol—the number of Btus expended among equipment manufacture, fertilizer, and diesel fuel required to produce a similar number of Btus of ethanol at the pump—vary widely. Recent work at UC Berkeley showed that producing ethanol from corn is more efficient than producing gasoline, and ethanol from other sources has far greater potential. California is considering requiring that all new cars sold in the state be capable of running on an 85 percent ethanol fuel; many vehicles are already available that can do that, and, at about $100, converting an engine to burn high ethanol content fuel is inexpensive. This is an area with prospects for significant impact on U.S. gasoline use. In Brazil, half the fuel needs of cars and trucks are met by ethanol produced from sugar cane and other crops.[9]

Solar

Certainly some technologies will become more economically viable as energy, in general, costs more, and solar is one of them—photovoltaic cells capture the energy the sun continues to pour on us even in our most wasteful moments. The production of photovoltaic cells worldwide has been growing gently since 1976, but starting in the late '90s, it has soared. The year 2003 saw a 32 percent jump from the year before worldwide, but, interestingly, production in the United States dropped 14 percent due to various causes, financial and otherwise, not least of which was the lack of dedicated funding from the government for the Million Solar Roof Initiative of 1997.

For more than ten years, Japan has been on the forefront of the solar movement, with its government offering major incentives. In 1992, a program paid for half of the installation costs of residential solar systems. That subsidy has been lowered since, but 168,000 residential systems were installed under the program, and their price has lowered dramatically.[10] Europe's solar use is growing fast, too, and Germany is the second largest market for the cells.

Costs of producing solar cells have been dropping and will continue to fall as production increases, unlike the production costs for many other energy sources. But solar systems remain more expensive than traditional ones, and until other energy rates go up considerably, this will likely remain the case.

Geothermal

Geothermal energy is in use in the United States today; geothermal heat systems keep hundreds of thousands of buildings at a comfortable temperature, and a small amount of electricity is generated using geothermal energy. In 2003, a little less than 0.4 percent of the total electrical sales came from geothermal sources. That is equivalent to about twice what was produced by wind power, and less than a tenth of what came from biomass. Thermal energy sufficient for electric generation is location specific—in the United States, Nevada appears to have the most potential—but to tap that source would take the same kind of geological exploration as does petroleum, which is hardly inexpensive. The resources are available to some degree throughout the country, and certainly increasing our use of geothermal energy for space heating could cut our dependence on hydrocarbons.

Hydrogen

Hydrogen is infinitely renewable, it is clean, and according to some, it promises a technology of great potential. Others characterize the attention given to hydrogen as simply hype. "It's all smoke and mirrors," says one chemical engineer, pointing out that there are only two reasonable sources of hydrogen: electrolyzing it from water (and what's the source of the electricity to do that?) or using natural gas as the raw material (and then you're back to the beginning).

The DOE held a conference in February 2002 entitled "A National Vision of America's Transition to a Hydrogen Economy"—the rest of the title sums up the problem—"to 2030 and Beyond." Despite President Bush's promise of commonplace hydrogen-based automobiles by 2020, most of the engineers slogging away at the problem seem to believe it will take more like thirty to fifty years to make the transition, even with a real commitment on the part of government and industry (a commitment that's presently lacking). Technologies for production, delivery, storage, and conversion must be developed—and a great deal of infrastructure is critical, as it is an unusual consumer who will want to play before the whole arena is set up and functioning. People are working on it, though.

The bottom line for renewables and alternative fuels is that Americans support the concept of going green, but they can be counted on to vote with their pocketbooks. Take the deregulated power offerings, for example; small is the number of customers who opt for environmentally friendly (but more expensive) electricity. Nuclear power may come back into favor eventually, but at this moment, without turning even more to coal (and its serious environmental threats), no alternatives to oil and gas are available that are of reasonable scale or economically feasible.

Probably the most sensible and responsible solution to our energy demands down the road would be to drastically improve conservation and efficiency of energy use, and there are many and varied ways to that end, but that is a political football well beyond the scope of this book.

Who's in Charge, Anyway?

If anyone's in charge, it's probably supposed to be the Federal Energy Regulatory Commission (FERC). Its website describes FERC as "an independent agency that regulates the interstate transmission of electricity, natural gas, and oil. FERC also reviews proposals to build liquefied natural gas terminals and interstate natural gas pipelines as well as licensing hydropower projects."[11] FERC's predecessor, the Federal Power Commission, was established in 1920; since that time, the number of employees has increased from one, an executive secretary, to more than twelve hundred, but as at the start, there are supposed to be five members of the commission itself. They are to have five-year, staggered terms, and no more than three are to be of one political party. As of February 2006, when literally dozens of LNG applications were before them, not to mention their myriad other responsibilities, only three members were sitting on the committee, all appointed by George W. Bush. One was a Democrat. There had been at least one vacancy for three years; the second came open in June 2005. Asked about the vacancies, a White House personnel staffer said, "We're working on it." He sounded as if he'd like to blame the Senate, as that body must approve a nomination, but he did admit that, as of then, the White House had put no names forward. However, within a few months they had filled the commission.

At this time, it is in the hands of this commission to approve siting of LNG import facilities. We will hear a lot more about FERC in the course of this book.

Natural Gas from the Beginning

THROUGHOUT history and long before anyone knew what it was, natural gas seeped to the surface and made itself noticed. Probably the best-known early Western example was in ancient Greece, on a plateau on Mount Parnassus, which from about 1400 BC through 381 AD was the site of the Apollonian Oracle at Delphi. Greek myth says that it was at this site that Apollo slew Python, the dragon snake that had taken over Parnassus. Henceforth, Apollo spoke from the temple through an oracle, always an older woman known as the Pythia, who sat on a tripod stool over an opening in the earth, breathing fumes said to be from the decaying body of Python.

Consulted by people on issues of all kinds, important and trivial alike, each priestess or oracle would be in a trancelike state as she made her prophecies and answered supplicants' questions. High priests would interpret the pronouncements, which were spoken in riddles—or at least in unclear words, as if intoxicated. Upon further payment by the answer-seeker, the responses would be added to, if not clarified.

During the time that the oracle was active, observers believed that the Pythia was given her powers of prophecy from the *pneuma*, the gas, vapor, or breath, which induced divine possession. Pliny, Plato, Strabo, and Plutarch all reported a sweet-smelling emission. For centuries, scholars accepted this explanation until it was debunked in 1900. No evidence remained of any chasm or emission, and even if there had been such, said English classicist Adolphe Paul Oppé, no natural gas could have caused a state that looked like spiritual possession. His pronouncement was accepted as truth for nearly a hundred years. At the end of the twentieth century, however, geologists found exposed fault faces crossing at the site of the oracle; analysis of the local water and rock structure and its makeup provided reasonable evidence that methane, ethane, and ethylene were present. A toxicologist joined the team of researchers and

21

found many parallels with today's teenagers who get high on glue or the fumes from paint thinner, as well as with earlier experiments on the anesthetic properties of ethylene.[12]

A number of temples of Apollo have been found to have been located over vents of odd gases (e.g., at Hierapolis, Dydyma, and Claros). Other historic shrines were also constructed on grounds that from today's point of view appear to have been the sites of natural gas seepage. In ancient days, these were noticed primarily when ignited by lightning. "Eternal" fires were the inspiration for the Zoroastrian fire worshippers, who built a number of temples in Azerbaijan. In Iran, such fires were seen as much as eight thousand years ago.

The decaying body of Python apparently spread throughout many parts of the earth; evidence of natural gas, along with oil and tar with which gas is often found, has been seen in seeps to the surface on all continents save Antarctica. (There is good reason to expect that Antarctica has hydrocarbon deposits as well. "Like all continents," I was told by one geologist, "it's been everywhere." Evidence exists of rain forest vegetation in the past, and fossilized remains of crocodile-like animals have been found. It might, however, take more energy to retrieve the oil and gas than would be recovered.)

Ancient Egyptians used liquid oil for lighting and as a medicine for wounds. The Greek name for the Dead Sea is Lake Asphaltites, because of the gooey lumps of tar that washed up on the lakeshore. For thousands of years, tar from the pits of La Brea, California, was used by Native Americans to waterproof boats and baskets. Natural gas was burned at least twenty-five hundred years ago when the Chinese used bamboo pipes to carry it from shallow wells to evaporate seawater to create potable water. French explorers in the early seventeenth century found Native Americans burning seeping gases around Lake Erie.

In 1821, near Lake Erie in Fredonia, New York, William Hart noticed bubbles coming to the surface of a creek and dug the first successful natural gas well there—all of twenty-seven feet deep. His efforts led to the first natural gas company, Fredonia Gas Light Company. For most of the rest of the century, gas was used only for lighting.

It was in the middle of the nineteenth century, in the backwoods of Pennsylvania, thousands of years after the earliest use of oil, that the modern oil industry got its start, and its time had almost—but not quite—arrived. Whale oil, previously the primary source of fuel for lighting, was disappearing fast and had become very expensive. The oil that could be derived from coal was also costly. During this time as well, researchers at Yale and elsewhere were learning how to purify and distill oil into several distinct products, each with their own potential uses.

The first commercial oil well was drilled near Titusville, Pennsylvania, also not too far from Lake Erie. In 1858, self-styled "Colonel" Edwin L. Drake thought to search below the surface near an oil spring. First digging by hand and then using existing drilling mechanisms, he persisted beyond what many considered to be reasonable, looking for oil. The next year, after his last backer had written instructing him to pack up his equipment and stop work but before he received that letter, at sixty-nine and a half feet below the surface, he found a significant flow of oil.

(We now know that his luck was extraordinary. Had Drake drilled a few dozen feet to one side or another, he would have entirely missed the pocket of sand that held oil. There were other oil-bearing formations more than a hundred feet lower in the earth and three hundred feet below that, but for economic reasons he never would have reached them even if his equipment could have managed it.[13] But luck is a large part of the story of oil and gas exploration.)

There was no shortage of oil in western Pennsylvania (only of whiskey barrels in which to store it),[14] and a boom economy grew. Life still wasn't simple for Colonel Drake, however. Who wanted his oil? How would he deliver it? How would he store it until he could ship it out? This last was a serious problem. In fact, one night the many barrels and wooden vats in his storage area exploded. Ultimately, Drake died a pauper, a fate awaiting many other would-be oil tycoons in following years.

Just as natural gas and oil became plentiful, the market for both came on hard times as great numbers of coal-fired plants started turning out electricity, a far simpler means of lighting cities and homes, and the price of petroleum plummeted. The boom was over, as was the land speculation that went with it. In one town, Pithole, Pennsylvania, a farm sold for $2 million in 1865—and at auction thirteen years later, it brought $4.37.[15]

The invention of the automobile changed everything for the oil business. By 1920, the owners of 23 million automobiles were demanding fuel, and oil-derived gasoline had been shown to be the most practical. Yet, there was still little demand for natural gas, which was often considered to be simply a nuisance in the search for oil. It wasn't until 1885, when Robert Bunsen invented his burner, that it became feasible to use natural gas for cooking or heating, though it was still difficult to transport the stuff. Much gas that was found was simply vented, burned, or abandoned.

With the development of pipelines to transport the commodity, however, natural gas became truly useful. One of the longer early pipelines was built from central Indiana to Chicago in 1891, but it wasn't terribly efficient. Although improvements were made steadily, it wasn't until after World War II that new technology and metallurgy made the pipeline transportation of natural gas reliable and efficient. Starting then, the vast network of pipelines that crisscross the country and surrounding waters grew rapidly. According to the U.S. Department of Transportation, there are now more than 1.4 million miles of gas pipeline in the United States, along with another two million miles of oil pipeline.

What Is Natural Gas, Anyway?

Natural gas is made up of hydrocarbon molecules, which are chemical compounds consisting only of carbon and hydrogen. Included in the hydrocarbon class are all the fossil fuels: coal, petroleum, and natural gas. There are several types of coal, and petroleum also varies in its composition, and can be refined into various products including gasoline, kerosene, asphalt, naphtha, paraffin wax, and fuel and lubricating oils. Similarly, natural gas varies considerably as it comes out of the ground.

Beyond their common fossil fuel base, natural gas has nothing to do with the "gas" we use in our automobiles, more properly called gasoline, which is refined from oil. Natural gas is colorless, shapeless, and odorless in its pure form. It is nontoxic, despite the Delphic experience; perhaps the psychedelic effects at Delphi came from some other associated material. As it comes out of the earth, natural gas is made up of several gases in varying ratios, depending on its particular source. Typically, the greatest part of natural gas is methane, from 70 to 90 percent, with the balance being the "heavies": ethane, propane, butane, and a few other denser hydrocarbons. Mixed in there may be a little bit of nitrogen, oxygen, and hydrogen sulfide (that's the one that stinks), and perhaps a trace of other, rare gases such as ethylene, and often water comes along for the ride. By the time it is delivered to a home in the United States, natural gas has been refined and is nearly pure methane; most of the other materials have largely been removed and an odorant has been added for safety. Propane, ethane, and butane are sold separately.

To many geologists, if not to the business people, oil and gas are simply the same thing and are considered in one breath. Often they are formed in the same manner and in proximity to one another.

Natural Gas from the Beginning

It takes millions of years and a long sequence of lucky coincidences to produce useable oil and gas. It starts with thousands of years' worth of accumulated organic material, dead plants and animals perhaps, from swamps or the ocean. This gets buried in mud and other sediment in vast amounts, and the more quickly it is buried, the better—overnight is wonderful, although exceedingly rare—because burial stops decomposition. The weight of the cover puts tremendous pressure on the original organic material, which is also exposed to very high temperatures either from deep in the earth or from nearby volcanic activity. The pressure and heat break down the carbon bonds in the organic material, and eventually, tens or even hundreds of millions of years later, you may have oil and gas.

It might seem hard to imagine the quantities of material necessary to cover the source material sufficiently to have such an impact—but consider that a flood can drop four or five inches of soil onto a river-bottom cornfield in a few hours. The Mississippi River dumps hundreds of millions of tons of soil and terrigenous detritus (material derived from the land, especially by erosion) into the Gulf of Mexico every year. Much of the Gulf has sedimentary deposits tens of thousands of feet deep. Add to that the normal movements of the earth, shifting plates and even continents, and the whole process becomes a little more comprehensible. Still, by the time oil being created today is fully cooked, I've been told, humans will no longer be around as a species.

Natural gas is also derived closer to and even on the earth's surface, biogenically, as tiny microorganisms in oxygen-free environments chemically break down organic matter into methane. These bacteria make their home in anaerobic conditions, such as swamps and rice paddies and landfills and the insides of cows, and they produce gas. There's much joking about cows' flatulence, but in fact, the belching of ruminants as they regurgitate their cuds for another chew produces significant quantities of methane. (We make faces when we think of the stench, but it's not the methane that smells so. Bacteria in bovine rumens also let off hydrogen sulfide, and that is certainly smelly stuff.) Methane makes up the second largest share of atmosphere-influencing, "greenhouse" gas produced today (behind only carbon dioxide, which comes from burning other fossil fuels). Although there is disagreement about how significant the methane in the atmosphere from domestic animals is, efforts are being made to minimize ruminants' burping by changing their feed. At the same time, there are efforts under way to capture the gas created by bacteria in manure by processing the animal waste in anaerobic methane digesters.

Methane from anaerobic digesters has been used to produce electricity in Europe for years, and increasing numbers of American farms are now installing them, thus decreasing the release of greenhouse gases both directly, by capturing the methane naturally emitted from manure, and indirectly, by lessening demand for electricity from coal- or oil-fired power plants.

And what about LNG? What exactly is it? Liquefied natural gas is simply natural gas that has been cooled to a temperature so low that it has liquefied, just as steam condenses to water when it gets below the boiling point. In the case of natural gas, the temperature required is about −260°F. Once liquefied, its volume is only one six-hundredth that of natural gas. That's like putting the volume of three basketballs into a golf ball. This is its advantage: in cryogenic form, natural gas can be transported over long distances by ship, allowing it to be delivered to places where pipelines are impossible, and it can also be stored in far smaller spaces. When warmed up, it turns back into natural gas. It has all the properties and usefulness of ordinary natural gas, with the added advantage that often a higher percentage of the heavier hydrocarbons, plus all the water and other impurities, have been removed. It is only for convenience of transport or storage that the gas is liquefied.

Finding Petroleum

Certainly the easiest way to find oil is to accidentally walk into a puddle of the stuff, but most of those easily accessed sources were discovered—and exhausted—early on. Drilling at random hasn't much chance of discovering anything, though there have always been people who tried drilling on their back forty, hoping to hit. Psychologists and animal trainers say there's nothing like intermittent reinforcement to encourage a behavior, and enough speculative wells turned up oil in the early days that a few entrepreneurs could always find backers for their next venture. Certainly there were vastly more "dry holes" with no oil to be seen, however. It is not entirely inappropriate that early Texas wildcatters have been called crackpot promoters and snake oil salesmen.

The term *wildcatter* is actually an honorable descriptor in the oil and gas business, and simply refers to a person engaged in speculative exploration in an area not known to be productive. But serious wildcatters are not flying blind; most likely, they are geologists who have specialized in this game, and have an understanding of both the land they're interested in and the habits of petroleum.

"Petroleum exploration is a very good way to lose a tremendous amount of money," says Frank Whitehead, a British petrophysicist with a knack for bringing things down to an understandable level. "There's no 'let's be lucky, let's drill a well or two.' Even if you drill with good knowledge, there's probably not going to be anything there—and you need to drill ten wells before you get a good idea." Of course, there are huge profits to be made, too, and the big players can afford more exploration to make more discoveries from which to make more money.

The easiest way to find petroleum is to look right next to an existing source, and even that can be unproductive. A reservoir may be huge—it might be a mile across, or in Saudi Arabia it might be hundreds of miles across—but it is not uniform across its width, and it does have limits. "You drill an eight-and-a-half-inch-diameter hole—you don't know what's two feet away," says Frank.

"If you don't like dirt and rocks, there's no reason to be a geologist," explains Mary Forristall, a paleontologist who worked in the petroleum arena a few years back. (Paleontologists, who study the history of life on earth, get involved in the oil and gas business because of their interest in fossil rock.) "Geologists like the out-of-doors—hiking, kayaking, all that. Therefore, we're ardent ecologists."

Robbie Gries, a wildcatter as well as a geologist, has been in the oil and gas business over thirty years, and, among other findings, is responsible for discovering a significant and productive biogenic basin in southern Colorado that her company is now developing. She has served as president of the American Association of Petroleum Geologists. "As a kid and in college, I knew I liked the earth—rocks, fossils, knowing how the earth was formed and made. Most geologists fall in love with the history of the earth and the way it functions." She reiterates Mary's concept. "We're the natural-born environmentalists. I grew up appreciating the earth, and I want always to appreciate it. While I was taking science courses at college, someone told me there's this course called Geology, and I said, 'Wow, I gotta try that!'"

By the beginning of the twentieth century, geologists had begun to understand some of the structures beneath the earth that could hold oil and gas. Once formed, both oil and gas flow upward as far as they're able. Being lighter than water, they compete for space in porous rocks such as sandstone and migrate until they hit an impermeable barrier—shale, or even salt, which when under enough pressure acts plastically, like Saran Wrap over a cantaloupe. Then the oil and gas stop, forming a reservoir. Often, this "trap" has been caused by the earth moving, folding or faulting, changing the arrangement of the subsurface rock and creating a lid atop a basin along which the oil and gas have been seeping. This trap is what must be found in order to harvest petroleum.

The most easily identified trap is an anticline, an arch-shaped fold in impermeable rock whose convex top can be seen from the surface. "That was when geologists had the most fun," says Robbie. "They'd ride a horse around Wyoming or Colorado or Texas and map anticlines. They were doing it in Mexico and Saudi Arabia, too."

But many anticlines didn't trap hydrocarbons, either because the source rock beneath wasn't rich enough or any oil and gas had migrated away before the anticlines were formed. "There are barren anticlines all over the world," says Robbie.

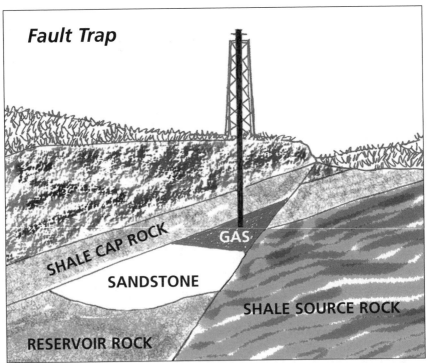

In a fault trap, layers of rock shift, making an impervious ceiling over gas-containing porous rock and trapping the gas beneath. KATHI PETERS

Beyond the obvious anticlines, there are more subtle traps, Robbie says, which are hard to predict. "You have to know about the deposition —where shorelines were, reefs, deep basin, the best source rocks, delta areas. It's putting together a lot of clues. Geologists used to go from creek bed to creek bed, looking at the rocks in outcrops, then go back to the office and add them all up. It does require a lot of patience. Old field geologists are incredible, they're so good," she says.

Searching nowadays usually begins with studying earlier records, though. Most of the major areas with petroleum deposits have been

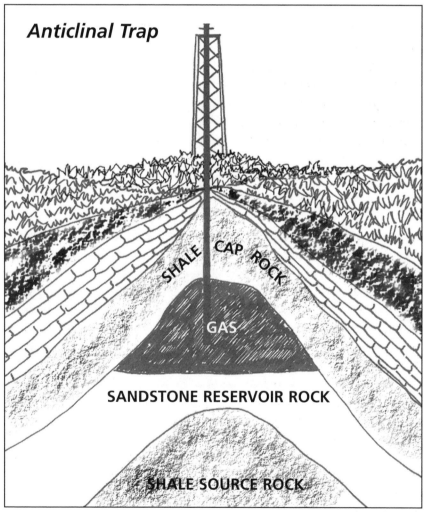

An anticlinal trap. As layers of rock become folded, domes are created that can trap pockets of gas. KATHI PETERS

identified. Robbie explains the sequence. "You look at records from other wells in the area, and try to extrapolate where hydrocarbons might have been missed. It takes a lot of imagination and creativity and knowledge of where rock might have been deposited." For people who like mystery games, she says, geology's the place to be. "Taking a few clues, you build a scenario, and you test it with lots of money and a drill bit."

"You find the ideal place where oil might have been made," explains Mary Forristall, "and then look all around the edge to find where it actually landed. With all the commotion the earth undergoes, it could go anywhere." Making the search difficult is that it can certainly migrate thousands of feet or several miles, and perhaps hundreds or even thousands of miles. And once the reservoir is identified, then the trick is to define its edges. Geologists again, analyzing rocks, joined by engineers.

It is important, though, that geologists get lost sometimes, says Mary. "Ants that don't get lost will starve," she explains. "An ant goes out, finds sugar, goes home, says, 'Hey, there's all this sugar out here,' and all the ants tramp out along the path to the sugar. But eventually it runs out, and if no one ever went off the track, they'd never find any more." Sometimes oil is found by accident. Drill for water, get salt—salt is a good indicator of possible petroleum, as it suggests that the site was once a sea, and it can itself act as a lid for the trap to hold oil and gas.

Salt formations can be disastrous, too, as Texaco found out in 1980 at Lake Peigneur, Louisiana, an area with a lot of salt domes. They wanted to drill a test hole in a lake on the edge of one of those domes, and had obtained permission from the Army Corps of Engineers just as they were supposed to. They had drilled down more than 1,200 feet when suddenly the rig started to tilt and the bit wouldn't pull out. In moments, a giant whirlpool developed in the middle of the lake, and the drill rig went down the hole, quickly followed by a second one and a tugboat, eleven barges, the barge loading dock, seventy acres of an island, parts of greenhouses, a house trailer, trucks, tractors, a parking lot, tons of mud, trees, and who knows what else. Two hundred million gallons of water disappeared.

"It looked like a lot of toys in a draining bathtub," said one onlooker. "They'd whip around, bobbing up and down, and then *bloop*, they'd disappear." A man and his nephew were fishing in a fourteen-foot aluminum boat; the water disappeared so fast that their boat was left stranded in the mud.

It turned out that the drill had broken into an active salt mine with eighty-foot-tall tunnels as wide as four-lane highways, supported by salt pillars. The water filled the tunnels, and when it hit the columns, they melted. While the Army Corps of Engineers *had* contacted Diamond Crystal, the owner of the mine, before granting permission to drill, some communication along the line didn't happen.

Amazingly, other than three dogs, no one died. The barges popped back up, but much of the other lost property was never seen again. The lawsuits seemed endless. "The best job at Texaco at that time was being a lawyer," concluded one reporter who covered the story.[16]

When you've identified an area of interest, the next step is to acquire oil and gas rights leases. You don't want to spend money for seismic work until you know you can drill. Then you lay out a seismic program. "Often it's a preliminary shoot to see if your idea is holding up. You may eliminate it right there. That happens more often than not," says Frank Whitehead.

It is analyzing this seismic work that has become Frank's particular expertise. "You make a big noise," he explains, "and the sound bounces off different layers." Just as a tennis ball bounces differently on concrete than it does on grass, so does sound react differently as it strikes varying materials below the earth's surface, and those differences are recorded. The data allows engineers to project the composition and shapes of the underground structures. There are two ways of generating the sound, with dynamite or by vibrations sent out by specialized large vehicles. The latter is less expensive than the former, but the surrounding terrain makes the most difference to the end cost. Obviously, where there are few roads and many trees, it will be difficult to use either system; helicopters can bring the equipment in and a crew on the ground sets it up, but that's extremely costly. On the other hand, going across a field of wheat in Kansas is simple. The explorers pay the farmer for his crop, of course. "That's the most valuable wheat ever grown," says Robbie Gries.

"If the clues are good," she says, "then you do a denser grid." The first shot—one line across the ground—might have only cost $6,000 a mile, but the more complex grid might be more like $30,000 a mile, and there are a lot of miles. They may run a line every twenty-five meters over an area as large as thirty or forty miles square, four or five miles deep into the earth. That makes for a tremendous volume of data—and Frank has written software to analyze it all. "It's a huge geophysics thing," he says.

"You spend four or five million, but it would cost twenty million to drill a well. Your success ratio goes up when you can see the structure under the ground." Still, says Frank, "you don't know what the rock is you're looking at. But then you drill a couple of tests, and the more you see, the more you understand the area."

It's still tricky to analyze what's been found when a test well is finally drilled. The stereotypical scene of the well gushing oil—black gold as it's called in the *Beverly Hillbillies* cliché—is extremely rare. In fact, usually nothing is visible to the naked eye, and sometimes even when petroleum is present, it is missed by the analysts.

Frank's first job, back in the late 1970s, was in Oklahoma. He ran the equipment that recorded information about what was in the hole. "Now it's computer controlled," he says, "but then there were banks of knobs—for an engineer that's wonderful, twiddling knobs. There was a big drum, an electric cable, and sensors, and you'd lower them down the hole. When we started, we put it on film eight inches wide and thirty or forty feet long—I'd analyze it by hand, and write my rough estimates on the logs—say, 'from here to here is sandstone. Looks like 20 percent porous, water saturation 40 percent, so I reckon the oil is 40 percent.' In the North Sea, with numbers like that they'd say, 'Go away—we'll look somewhere else,' but in Oklahoma, back then, they'd have thought 10 or 12 percent porosity would be good." Sometimes they'd be fooled, and they might find something useless like carbon dioxide making up the non-water balance. A significant proportion of carbon dioxide will make the well less appealing, as carbon dioxide attacks steel and destroys equipment.

In the 1980s, Whitehead was working for one of the major oil companies in Pakistan. After ten or twelve wells they'd had only one discovery, and it wasn't very big. But after quite a few more test drillings, they hit another, and then they worked out where the petroleum was trapped and how, and did much better.

"Nowadays, electronic equipment goes down the hole and records data every six inches—basic properties of the rock, like resistivity (how well electricity flows through it), hydrogen index, electron density, natural radiation, and other things, and people like me turn the recordings into things that make sense."

Still, says Frank, "The best geologist is a lucky one."

Robbie Gries has had her share of luck, and loves the game. "The most fun is the hunt, the mystery. And geologists are fun to work with." Their outlook is down-to-earth, dealing with millions of years. "Everyday things may get you down, but then you've got that perspective."

From By-product to Commodity

Many wells produce both oil and gas. "If the primary source of the black muck organic material was marine—algae, fish poop, and so on—the more oil there will be in the product. If it was land-based, as comes down a river, it will have a greater share of gas," explains Mary Forristall.

But gas was merely a nuisance in the early days, and the only time spent searching for it was in order to avoid it. It's much easier to capture and move oil. "When gas was worth ten cents a thousand cubic feet and pipelines were required, it was a nuisance," says Robbie. They used to simply burn the gas off until the 1930s, when environmental and economic considerations put a stop to it in many places. More recently, "they'd plug the well and move on." In Nigeria and the Middle East, there is still firing that can be seen from the space shuttle at night. (Nigeria has one of the largest concentrations of proven natural gas—and potential revenue therefrom—but as of 2005, it was still flaring over 40 percent of what it was then producing—about a quarter of all the gas flared globally.)[17] That we may have already wasted half of the natural gas that once was available is certainly unfortunate.

By 2000, natural gas had risen to more than three and a half dollars per million Btu at wellhead, and now, with prices regularly above six dollars and sometimes exceeding ten and even fifteen dollars, it's worth looking for.

Once natural gas has been brought under man's control, it is not always needed immediately. It can be stored indefinitely, either liquefied, in tanks, or, as is more common throughout most of the United States, in natural underground storage facilities. These are usually depleted natural gas or oil fields close to consumption centers, although natural aquifers and salt caverns also are used. They most often are owned and operated by interstate pipeline companies, local distribution companies, or, to a much lesser degree, independent storage providers.

The Hydrocarbon Value Chain

"As you follow the whole hydrocarbon chain, the personalities of the people involved change," observes Jed Watkins, facility engineer at the only American export terminal, in Kenai, Alaska. "Drillers wear tall boots and have egos they haul in semis, and to them 'close' means within a couple of orders of magnitude. The other end, at the refinery, where they're making splits between molecules, they're talking parts-per-million specs. Anal retentive. You pick your piece of the chain. I'm a midstream guy—tank farm: plus or minus 5 percent works for me."

Watkins was speaking about the petroleum value chain in general. The value chain for liquefied natural gas, in particular, has five links. The first, "upstream," as it is called in the oil and gas business, is exploration and production of natural gas. This is potentially the most profitable aspect of the business, but requires the largest capital outlay and certainly is the riskiest—it is common enough to lose millions of dollars on one upstream effort. It's no playing field for minor leaguers. The second phase, liquefaction, has allowed for the profitable sale of gas with no previous market, and also represents a significant capital outlay, as does the third, shipping. Regasification has the lowest price of entry, at a half or seven-tenths of a billion dollars. Each of the last four stages is more likely to make 10 percent on investment, plus or minus. Headed "downstream," the product leaves the regasification plant by pipeline, the fifth link, to the end market. In the next chapter we will look at some representative companies involved in the natural gas distribution network.

The LNG "value chain." KATHI PETERS

Aspects of the LNG Distribution Business in the United States

THE greatest share of the natural gas used in the United States is consumed in the Northeast and Midwest, and most of it comes in from the gas fields by pipeline. During the last few decades, a small amount has been imported by ship in the form of LNG. In 2005, there were five operating LNG import terminals in the continental United States. The first was the Suez plant in Everett, Massachusetts, which received its first shipment in 1971. Thanks to the premium price natural gas commands at the end-of-the-line New England market, it has operated ever since, with the exception of only a few weeks after September 11, 2001. Elba Island, Georgia, received deliveries from 1972 to 1980 and again starting in 1989; Cove Point, Maryland, operated from 1978 to 1980 and started up again in 2003; during the 1980s Lake Charles, Louisiana, took LNG only infrequently. (EcoEléctrica, a new import terminal in Puerto Rico, became operational in 2000. All its natural gas went to electrical generation originally, but it is the intent of the government to "gasify" the greater part of the island eventually. Puerto Rican energy figures are not included in the EIA figures listed in this book.)

Another, very different import facility received its first delivery in 2005: Excelerate Energy's Gulf Gateway Energy Bridge off Louisiana (more about this offshore terminal in the next chapter). These facilities together have the capability of sending 4.215 billion cubic feet (Bcf) of natural gas per day into the national distribution system, or a total of a little more than a trillion and a half cubic feet each year. In 2003, not even one trillion cubic feet of gas was produced in these facilities—only 1 percent of our annual natural gas usage came into the country in liquid form. Projections made in 2005 suggest that this amount would more than double in 2006, and increase tenfold by 2030; by then it would be more than triple the amount coming in by pipeline.[18]

Many people who either are or would like to be in the natural gas business believe in the strength of the future for LNG; the rush to develop is on. In 2005, fourteen projects in the United States (including pipelines from the Bahamas) had been given final approval by FERC or the U.S. Maritime Administration (MARAD), with an overall capacity totaling 6.1 Tcf/year. Another five were approved for construction in Canada and Mexico, with the potential of adding another 5.1 Tcf/year. Not all of these will be built, of course. Some have been officially canceled; others just aren't happening yet and who knows if they ever will. Still, as of the end of 2005, twenty more projects were formally proposed to FERC and the USCG—adding up to 9.2 Tcf/year—and ten more were being discussed in the United States (another 2.15 Tcf/year). Another dozen are also under consideration in Mexico, the Bahamas, and Canada.

Simple arithmetic would show that far more facilities are being dreamed of than will be needed, even if considerably more demand were to develop than has been predicted. How many applications will be approved no one can say, and how many approved projects will be constructed is another issue entirely, and one that comes down to economics. Obviously every prospective importer has to consider the whole value chain. Some companies are involved at all levels, like British Petroleum; they own the natural gas upstream, they pull it out of the ground, and they own each stage heading downstream. They liquefy it (BP has interests in four liquefaction projects) and ship it (they even own three tankers themselves). They own shares of vaporization facilities in Europe that send natural gas to end markets there, and in this country they have long-term contracts to supply the Cove Point facility. They are planning regasification plants of their own in the United States and other parts of the world. Suez Energy, on the other hand, which has interests in power generation, is not involved in the exploration or production end of things. They have partial ownership of one liquefaction train facility in Trinidad, they own one ship and have long leases on three others, and they regasify and sell downstream. There are also companies who hope only for an interest in vaporization, such as Downeast LNG in Maine. Cheniere Energy started in that manner. More about them later in this chapter.

There is a considerable lead time on creating a land-based import terminal, with a minimum of a couple of years in the planning and application phase and another three in construction. Some of the offshore projects have a shorter completion time. Each project faces much scrutiny, as indeed it should, and much aye- or nay-saying without any scrutiny whatever, which is far less productive.

Pipelines

All over the world, pipelines carry natural gas from place to place. They are the simplest way to transport gaseous matter. They vary in diameter from three inches to four feet, and form an arterial system with little pipelines in production areas feeding into larger ones that lead to bigger ones yet to carry product long distances, and then, capillary-like, smaller and smaller ones fan out again within market areas. In 2002, according to the Department of Transportation, there were 1,411,381 miles of gas pipeline in the United States, excluding service pipe. A little over 22,000 miles were gathering lines from production sources, and some 309,500 miles were large transmission lines carrying the gas from the area of its harvest to local distributors, but the vast majority distributed the gas to more than 60 million industrial, commercial, and residential customers.[19]

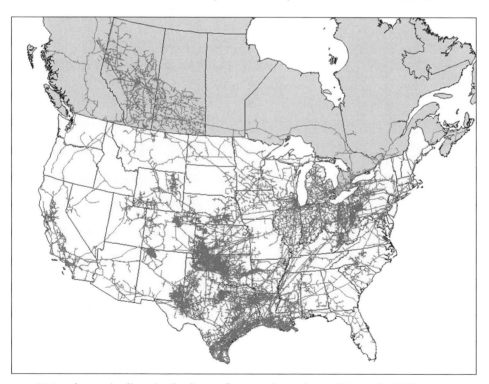

Natural gas pipelines in the lower forty-eight states and Canada, 2002.
COURTESY OF PENNWELL MAPSEARCH

Maritimes & Northeast Pipeline

As an example, let's take a look at one transmission line. The 800-mile transmission system belonging to Maritimes & Northeast Pipeline (M&N) was built to deliver natural gas that comes ashore from ExxonMobil's production fields on the Scotian Shelf near Sable Island, 185 miles east of Nova Scotia. M&N transports the gas to Massachusetts, crossing Nova Scotia, New Brunswick, Maine, New Hampshire, and Massachusetts along the way. In the Boston area, it connects with other pipelines running west and south all the way to the Gulf of Mexico, but its primary purpose is to connect with local distributors.

The company owns no petroleum itself, nor extraction or processing facilities. It simply provides transportation. But since it belongs 77.53 percent to Duke Energy, with lesser ownership by Emera, Inc. and ExxonMobil, its superiors are well entrenched in the oil and gas business. (This is very typical of the oil and gas business. Each little piece has its own corporate structure but comes under the umbrella of one or more larger entities, and it's all very convoluted and confusing.) M&N's customers are actually the buyers of the natural gas; in effect, the company is the trucker that picks up product at the factory and delivers it to purchasers. The customers include, for example, utilities such as KeySpan Energy, the large natural gas distribution company in Massachusetts and vicinity.

"We are an open-access pipeline," explains Marylee Hanley, the company's manager of government and public affairs in the United States. By FERC rules, it is the obligation of the owner of a pipeline to create more capacity when there is a call for it. After all sorts of scrutiny, any company or utility may connect to and be serviced by the pipeline, either to send gas to customers in other parts of the country or to purchase gas for local sale. In this country, to build or expand a line, the company files an application with the FERC. Their proposed routes take terrain and many other issues into consideration. As much as possible, they try to build within existing utility corridors, such as those of electric transmission lines. FERC, after consulting the Environmental PProtection Agency (EPA), the Army Corps of Engineers, and another ten or twelve agencies—as well as seeking input from any interested people, must conclude that it is "to the good of the public" before any building or expansion can be undertaken. FERC alone has the ultimate say in whether a project goes forward, and in just where an expansion will take place. In fact, says Marylee, FERC changed about half of the right-of-way for M&N's original main line from the proposal the company had offered (much of it in very minor ways).

In the state of Maine, M&N has 275 miles of pipeline and two compressor stations to move the natural gas along. Dave Peppard was a selectman in Richmond when the pipeline was proposed, and now is town manager there. Although there was a group opposed to the coming of the pipeline, concerned about devaluing property and the perception of danger, Dave says it has been a huge benefit to all the taxpayers. The company listened to the town about placement of the right-of-way, moving it to run along property lines when possible. "I can't say enough good about them," says Dave. The company pays taxes on their holdings in each municipality; Richmond, where one of the compressor stations is located, receives a total of $610,000 a year. In the little town of Morrill, where the pipeline right-of-way is within Central Maine Power's electric line corridor, M&N paid more than $44,000 toward the $609,000 locally raised share of the town's budget for 2005.

M&N has announced expansion due to the construction of an import terminal in the Maritime Provinces of Canada that is planned to start operation in 2008. Six new compressor stations (Richmond is due only for an expansion, not a new compressor, but it will presumably benefit financially from the greater capacity) and 146 miles of new pipeline "looping," with construction of a second pipeline parallel to an existing one to increase capacity, are proposed along the U.S. portion of the line. "Sure, we'd absolutely love to have more equipment in town," says Dave, but there's no plan to increase the infrastructure in Richmond at this time.

In Canada, new lateral lines must be constructed to the new terminal, and compressor stations will be added along the way there, too. M&N has assured the developers proposing LNG terminals on Passamaquoddy Bay, which would need to connect up with the pipeline to transport their product to market, that they will have capacity for gas imported to down east Maine as well. Part of Marylee's job is meeting with the various governmental officials, starting with the governor and continuing on down to the selectmen in each of the towns that the pipeline would cross.

The existing steel pipeline was constructed in 1999, preparatory to the opening of the Sable Island gas field in 2000, and is completely underground. It consists of thirty- and twenty-four-inch diameter carbon steel pipe about an inch in thickness and coated to protect it from moisture. It is laid in a trench six feet under surface level and covered with soil. "It is electronically monitored twenty-four hours a day," says Marylee. "There are valves every few miles, and if the gas control folks saw a pressure drop, within seconds, they would shut the valves down north and south of the problem and isolate the section." If any gas were to escape, of course, it would simply evaporate into the atmosphere anyway, unless it happened to ignite. Pipeline explosions do occur occasionally, but considerably less often than they used to, due to strict regulations and standards for maintenance.

At regular intervals, employees walk the right-of-way, and they make "pig runs," inserting a robotic Smart Tool, or "pig," into the pipeline. "It's like a scrub brush," Hanley says. It has sensors that read the wall of the pipeline and identify areas with corrosion or damage. "We Smart Tool it and look for anomalies, and if we find any, we go in and fix them." They close off sections and replace them when needed, though that hasn't been required yet on the M&N pipeline. She knows how it works from working for Duke Energy's Algonquin line from Massachusetts to New Jersey, up and running since 1949. (She went to work for Algonquin straight out of college, with the intention of getting a year's experience and leaving. She has been there ever since.)

Marylee does say that since September 11, they don't discuss publicly all the precautions they take. But there's no question in her mind that "We are very, very serious about safety." A little plane, its wings clearly labeled PIPELINE, flies low along the right-of-way every day in both directions. It used to be that the pilot would be looking for yellowing leaves, which signaled a gas leak, but now that the electronic monitoring is more accurate and timely in its information, the mission of the plane is still to look for yellow—or blue, or purple, or whatever color an excavator or bulldozer might be. "We're mostly looking for encroachment," Marylee explains, "forestry clear-cutting, or someone building a swimming pool or a foundation." The most common threat to pipelines today is from construction activities—nearly half of the serious incidents involving pipelines each year are related to "damage from outside force," most due to third-party excavation.[20] Although abutting landowners often have certain rights to the surface over the pipeline, such as pasturing or haying, they must check in with the authorities before they dig near the line. Every state has set up a program with a single phone number for people to call before excavating near a pipeline; the pipeline company is then alerted and may mark the area or send people to watch over the digging.

One morning Penson Bartlett, a heavy-equipment operator from North Searsmont, Maine, was just starting to dig a cellar hole on property adjacent to the M&N pipeline when he saw the plane fly overhead, circling once. Within twenty minutes, representatives of the pipeline drove up with a photograph in their hands of his truck, his name and phone number clearly showing on the door. There was no difficulty with where he was working, but they wanted to make sure.

High-tech systems are available for use in pipeline reconnaissance planes, too. These devices can identify leakage by using various imaging and other remote sensing techniques. Whether these systems are in use on a particular pipeline most likely isn't information the pipeline people want to disseminate, but they're being improved all the time.

But pipelines can't be built just any old where—there is an underwater mountain ridge nearly the whole length of the Atlantic Ocean, for instance, caused by volcanic activity and magma rising through rift zones where tectonic plates are still pulling apart. In fact, the Atlantic grows at the same rate as do fingernails, some 2.5 centimeters a year.[21] And because transporting gaseous matter in containers is extremely inefficient, that's where liquefying natural gas comes in.

Kenai: Liquefaction and Exportation

"We like to say we were the first commercially successful plant," says Jed Watkins, the chief plant engineer at ConocoPhillips's Kenai LNG liquefaction and export facility in Alaska. "Algeria was first, but they had problems, largely because they were in Algeria." The political scene there has not always contributed to smooth business dealings overseas.

As of 2005, there were twenty-nine liquefaction and exportation facilities worldwide, and more were under construction; Kenai is the only one in North America. The plant made its first shipment to Tokyo in 1969, and has been continuously sending ships back and forth to this day, with a load leaving Kenai every nine or ten days. Well over a thousand voyages have been made. In the earliest days, they provided 100 percent of Japan's LNG—now, with increased LNG production around the world and increased demand in Japan, their contribution is only 2.5 percent of the total.

Natural gas comes into the facility from the gas fields of the Cook Inlet and is sent through a closed-loop refrigerating system fired by propane, ethylene, and methane stripped out of the raw product before liquefaction. It works like a giant, super-cold household refrigerator. Once liquefied, the product is stored in big tanks, from which ships load and head off for Japan. "The whole process is neat," says Jed. "Very clean. Well-defined steps." Nearly the first liquefaction plant constructed, the Kenai facility is perhaps not the most modern, but it's well maintained and still functions efficiently, and similar proprietary technology is still being installed in new facilities around the world.

The two Swedish-flagged ships on this run are small, at only 89,880 cubic meter capacity. (The typical carrier is now 125,000 cubic meters or better, and new ships are being constructed up to a capacity of 250,000 cubic meters.) A thousand-yard security zone is established around the tankers as they come and go, but the self-docking ships are not escorted either by Coast Guard or by tugboats. Registered fishing vessels are allowed within that area, and the tanker masters and fishing vessels report infringements to the authorities for enforcement. (If national threat levels were elevated, more security would be in place, the Coast Guard says.) Pilots do come aboard for the approach, but the ships' masters are familiar with both the waters and the other vessels sharing them. The schedule is not as widely publicized as it used to be, though some of the ships announce their coming for the benefit of the drift-net fishermen, who need a couple of hours to collect their gear. There has been some impact on fishing, but the company has reimbursed fishermen for lost gear. Once a carrier is tied up at the jetty, the tension on the mooring lines is monitored electronically, to check for ice buildup as well as any other irregularity.

A fire has never been reported on the Kenai site, but there are established drills so everyone would be prepared if a leak or a fire were to occur. They have performed their own testing of vapor cloud formation, finding that the cloud is no wider than the spill and dissipates quickly. Test fires ignited in the limited spill situations they've set up quickly burn themselves out. Their fire department has twenty-five full-time firefighters and thirteen volunteers, all trained by Texas A&M faculty. The company may not feel it's necessary to have such a large fire department on their account, but it can't hurt.

The only environmental issues anyone talks about are seawater ballast dumping, with the concurrent potential of introducing foreign species (which would not be an issue at an import facility, as the rest of the North American plants will be), and the temperature of the discharged processing water. There is no effort made to limit the impact of lighting, as the plant is in an industrial area, with a fertilizer manufacturer, a refinery, and a gas-to-liquids facility nearby. In general, the plant gets nothing but high marks from locals. While it only employs forty-two full-time people and ten contract staff, that is but a small part of the full local economic impact. One oil services provider bills Kenai $6 million a year. The facility is a significant part of its community of seven thousand people.[22]

Import Terminals: Distrigas, in Massachusetts

All told, about a fifth of New England's total energy comes from natural gas, with Massachusetts using a full half of that. The region receives between 80 and 85 percent of its natural gas by pipeline from Canada or the Gulf of Mexico. Julie Vitek is communications director at Distrigas, a division of Suez Energy and natural gas provider to Greater Boston. She explains, "We're at the end of the interstate pipeline systems—we're near the end of the Canadian system as well as the Gulf's." That places the region at risk of coming up short. "Imagine that it's a long drinking straw, and everyone along the way takes a sip as it goes by." New England is beyond New York and New Jersey and a lot of serious gas users. "The supply is pretty tight now," Julie says. The rest of New England's gas needs are met from the LNG that passes through the Distrigas Terminal in

Assisted by tugs, the carrier Matthew *backs toward the Distrigas plant in Everett, Massachusetts. The two largest tanks on the right are Distrigas's storage tanks. Behind the facility, toward the left, is the gas-fired Mystic Power Station. The berthed oil tanker is at the Exxon dock; Distrigas's dock is upriver from that.* COURTESY OF DISTRIGAS

Everett. As demand for natural gas builds, there will be more of a role for LNG in the region.

Sixty-three times in 2005, an LNG ship passed through Boston Harbor on its way to the Suez Distrigas LNG import and regasification facility on the Mystic River in Everett. Located on a site formerly occupied by a plant that manufactured gas from coal, the terminal was initially constructed as an LNG peak-shaving facility (peak shaving explained in more detail beginning on p. 58). Converted, it became the first LNG import terminal in the United States and received its first shipment of LNG in 1971. With only a few short interruptions, it has operated continuously ever since, receiving some 700 cargoes up through 2005.

Although there were three other import facilities built in the United States not too long after Everett's, natural gas prices were so low in the 1980s and 1990s that two of them were mothballed and the third was only kept operating at a minimal level. New England isn't unique with its cold winters, but unlike much of the rest of the country, it has no natural gas supplies of its own, and no underground geologic formations in which to store natural gas. LNG continued to make economic sense at Everett when it didn't elsewhere. In the late 1990s, Cabot Corporation, then owners of Distrigas, believed that imported natural gas had a future. With four other companies, they invested in the development of an LNG liquefaction plant on the island of Trinidad, in the Caribbean. That facility came on-line in 1999. Gas prices were still low then, but the unique New England situation allowed LNG to be economic even if not brilliantly profitable at first.

"We were a couple of years ahead," says Frank Katulak, senior vice president of operations at Distrigas, "but we weren't surprised to see the increase in demand." Just as domestic natural gas sources were leveling out, the number of gas-fired electrical generation plants was increasing, creating a need for more gas and making gas prices rise. LNG was again profitable. The mothballed LNG terminals went back into business, and deliveries to Distrigas increased substantially. In addition, the Trinidad plant has increased its production capability several times since opening.

Distrigas serves customers on a daily basis, and with weekly, monthly, seasonal, annual, and multiyear contracts. They provide natural gas, both liquid and vapor, to a number of other customers in the northeastern United States, including the (presently) forty-six utility-owned plants in New England that store LNG for use on cold days. Forty to sixty trucks a day leave the plant with LNG headed to satellite and peak-shaving facilities around the area, 12,000 loads a year, Frank says.

The Everett terminal is well situated for the New England market, says Julie. "We are putting gas right in the market that uses it," she explains. They have direct access from two interstate pipeline systems, Tennessee Gas and Algonquin, as well as to the local KeySpan distribution pipelines. They also have a direct line to the 1,550 megawatt Mystic Station electric power plant next door, their largest customer. Originally only oil-fired, the power plant has been in operation for years, but adding natural gas to their capabilities in late 2002 increased production significantly. Today, the plant operates on natural gas as much as it can, because it is both more efficient and less polluting than oil. Mystic Station can provide about three-quarters of the electricity for Greater Boston, an area with two and a half million residents that includes Boston proper and the adjacent cities and towns.

"The plant is not very labor-intensive," Frank says. Although the plant runs seven days a week, twenty-four hours a day, only twenty-eight people total actually work hands-on, and the rest of the employees are support. Frank has been at Distrigas for seven years, and during most of that time, they've been looking for qualified people to work in the plant. It's not necessarily that there are openings, but it's not easy to find the right experience. "We want to hire people with some process-plant experience—power plant operators, oil refineries, or LNG carriers." There just aren't many. "If we find one, we'll take a qualified person in. If we were

Maintenance workers at Distrigas. COURTESY OF DISTRIGAS

in Louisiana or Texas, we'd just have to put a sign out at the door and we'd be inundated."

Training for employees follows on two tracks. The federal regulations specify basic training for anyone working with LNG, and Distrigas fulfills those requirements but adds others of its own. A new employee starts as a C operator, and there are four levels altogether. Employees learn and study as they're on the job, with classroom work and manuals to read. "There's a whole list of things to get 'signed off' on," Frank explains. They work with someone else and they do their homework and do each particular task under supervision, and when they understand and show they can perform a specific job, that item gets checked off. They must get fifty things signed off on altogether, and then they're allowed to take written and hands-on tests, and then they move up a level. Some of these tasks are things they'd have to do on a regular basis, and others are for abnormal situations, "but the test is not a 'Gotcha!' kind of thing." Typically, it might take someone five years to get from C to A level, where the employee is capable of running the shift alone and handling every operation. They want three operators and one supervisor, the fourth level, Lead, on each shift. "It's a long learning curve," says Frank.

There are also three categories of maintenance people: mechanics, who work on pumps and compressors and so on, electricians, and instrument technicians, who work on controls. Some overlap exists between the last two. It's easier to hire people for these positions; a licensed electrician may have experience with high-voltage equipment at a power plant or in any number of other industries, for example. Still, they all are taught the basics of LNG safety as required by the federal regulations and the company's own rules.

Many *dos* and *don'ts* need to be learned. Do, before you do any work, make sure that the area has been checked for gas, that there are no leaks, and that the equipment is safe to work on, electric breakers are locked out and tagged. Do wear uniforms, hard hats, safety glasses, and shoes. Don't, if a ship is docked, go aboard until cleared by customs, immigration, the Coast Guard, and the captain. Don't just walk up to a leak of LNG from a flange and tighten the bolts.

Flanges do leak occasionally. In Frank's seven years, there have been no leaks other than from flanges, and those maybe only once every six months to a year. It's easy to understand why they might leak, when you think about the temperature of LNG. Imagine a piece of metal at ambient temperature. When it comes into contact with LNG at −260°F, it will contract. The 1,500-foot-long pipe from the loading dock to the tanks shrinks several feet, Frank says. (The ends don't move; there are expansion joints along the way that accommodate the changes.) At a flange, where two pipes are joined together, a bolt might not contract quite as fast as the flanges. If the gasket isn't just perfect, or the flange bolts aren't torqued properly, it might lead to a leak. It would simply be a drip, accompanied by a little cloud where the water vapor in the air is condensed. The drip is likely to vaporize even before it hits the ground; if it does hit, all you will see is a little frost spot. But you don't just run in and tighten the bolt. You get your safety equipment, because if a drip contacts bare skin, it could cause frostbite. You wear rain gear like a fisherman's, a plastic face shield, and gloves, and then you tighten the bolts, and not at random, but in a prescribed pattern. "A lot of what we do here is not rocket science," says Frank. "It's common sense. It's not hard, but it's peculiar to LNG."

The real complexity of the business primarily lies in managing the ships in and out and supplying the gas to the customers.

There's a regular routine when a ship comes in. Everyone knows the drill. Longshoremen take care of tying up the vessel while the Distrigas onshore team is preparing to receive the product. At some terminals, they pump LNG out from a shoreside tank through the liquid lines to precool them before unloading a ship. At others, the lines are kept cold at all times. Three different communications systems are used to be sure the ship and the terminal can talk with one another. Aboard the vessel, the crew is preparing for cargo discharge. When the ship has cleared customs and immigration, the cargo officer spends a few minutes with the receiving team, talking about the schedule, how fast to unload, and so on.

A lot of money is at stake, millions of dollars for each shipload, so measuring the cargo is an important step. A team of three—an independent cargo surveyor, someone from Distrigas, and the ship's cargo officer—read the level gauges in the ship's control room and sometimes out on deck, too. The team makes note of the list and trim of the vessel, the pressure and temperature of the product in the tanks, and then they go off to make their calculations of the actual quantity of LNG that the ship arrived with.

These gauges are critical to everyone, as they are a key part of the inventory system. Maintaining and fixing the gauges is the specialty of Hector Cortina of CHI Engineering Services in Portsmouth, NH. "There is

Piping connections to cargo tanks on tanker during discharge at Distrigas.
COURTESY OF DISTRIGAS

more than one system," Hector explains, "but when one fails, they call the representative." That's him. They have to have redundant operative gauges, or they might get to Trinidad and not be able to load one or more tanks, which would cost everyone money. "It all has to match," Hector says. "When they get to Boston, they have to know exactly how many cubic feet left Trinidad, and how much gets to Boston." Some boil-off has been used as fuel along the way. "It goes into the billing process."

The tanks on the ship are all joined together through a grid of valves and insulated piping on deck leading to a manifold area amidships, where the connections to land are made. The receiving terminal has the information about the manifold arrangement on the particular ship—how high above the water, how far inboard, how far apart the manifolds are—and they're ready to hook up the loading arms once the cargo survey is taken. At Distrigas, as is typical, there are four LNG liquid loading arms, which feed into large lines, twenty-four inches in diameter. At some facilities, those lines are as much as forty-eight inches in diameter.

Each cargo tank has its own pumps—typically two per tank, in case one were to fail. The pumps move the LNG through the unloading line and up to the top of the receiving storage tanks. Although most of the piping on deck and on shoreside tanks is insulated, the loading arms are not. They build up ice that itself acts as insulation during the unloading. Pumping off starts slowly and from only one of the ship's tanks, but within an hour they're going full bore.

It takes about eighteen hours to unload at Distrigas, all told. Nobody's doing much during that time. Crewmembers and terminal personnel monitor the controls and make periodic inspection tours. Every so often, people change shifts, relieving each other. A few crewmembers take shore leave. Supplies come out and are loaded onto the ship. That's it.

At the end, the procedure reverses. Unloading slows down and then stops. The cargo surveyors come aboard again and measure the LNG remaining (a small amount is always left to keep the tanks cold for the next load), then go off to do their arithmetic. The loading arms are purged of LNG with inert nitrogen and disconnected. "It's not much different from unloading any other petroleum product—just the temperature," says Neal Frangesh, a mechanical engineer whose specializations include cryogenic applications. The time from when the ship arrives at the dock until her lines are off and she's on her way again is just about twenty-four hours.

At Distrigas, the ship unloading line carries the product 1,500 feet ashore to one of two tanks. Modern LNG tanks have over-the-top pumping, with no piping connections out the bottom of the tank, although many tanks have internal piping to permit discharging the LNG into the top or bottom of the tank, depending on the composition and density of the particular load. If the LNG being unloaded is heavier than what is in the tank already, they'd send it into the bottom of the tank in order to prevent a phenomenon known as rollover. Rollover occurs when layers of LNG of different densities try to rearrange themselves; as the different layers meet, some of the liquid might vaporize and cause pressure build-up in the tank, which, in turn, could lead to a large LNG venting or even the structural disintegration of a tank. The makeup of each tanker load can be pretty well estimated by the port of origin—most LNG from one location is consistent—but a certificate of quality from actual sampling at the loading plant accompanies the cargo and is among the things the surveyors check. The product within the tank is carefully monitored, too.

The next phase is vaporization, or regasification. Baseload terminals vaporize most of the time and send natural gas out to the pipeline system almost continuously. Submerged pumps within the storage tanks are used to push the LNG out of the tank and then through insulated piping to the vaporization facility, not too far away. There are three or four ways of vaporizing, Frangesh explains, and they use two different methods at Distrigas. Submerged combustion uses stainless steel coils running through pools of heated water. The other method uses hot water from boilers and heat exchangers. Close to 2 percent of the natural gas thus vaporized is used to fuel the system. Odorant is added to the gas going into the KeySpan distribution lines, which leaves at a relatively low pressure. The gas headed to the power plant next door or to the Algonquin pipeline is sent out at higher pressure, up to 700 pounds per square inch. And there's always LNG in the Distrigas tanks for trucks headed to satellite facilities around the Northeast.

Escorting LNG Carriers into the Everett Plant

It's not that the LNG ships that come into Boston Harbor are particularly big. "At a little over nine hundred feet, they're nothing extraordinary," says Jake Tibbetts, whose Boston Towing and Transportation has the contract for tugboat assist for the ships. "They're the same size as container ships." It's the security that makes the job unusual. Presently, Suez Energy brings a tanker in about every five to seven days. The new ships, the *Berge Boston* and the *Berge Everett*, each have a capacity of 135,000 cubic meters of LNG. The older *Matthew* is only a little smaller. Every time one approaches Boston Harbor, a veritable flotilla of vessels of varying sizes runs out to meet it. Most of those boats are simply to protect it, and everyone hopes they don't have to do anything but run alongside. Jake's tugboats actually have to work.

Fog is the only weather condition that prevents the entry of an LNG

Tugs assist the carrier Berge Boston *under the Tobin Bridge and toward the berth at Distrigas.* COURTESY OF JAKE TIBBETTS

tanker into Boston Harbor—the Coast Guard requires two miles' visibility. Day or night, it is the tide that determines the time of passage. Four tugs meet the ship at the entrance to the harbor, at Deer Island, seven and a half miles from where she's going. The deckhand on one of the two tractor tugs, the *Freedom* or the *Liberty*, throws a line up to the men on the stern of the ship, which is still moving at eight knots. The security rules say that a tractor tug must be tethered to the carrier whenever it's in transit through the harbor, and the other three tugs must be in attendance. (Only *Liberty*, *Freedom*, and one additional tug are required to take the empty tanker back out to the open sea.) The tractors were purpose-built for this contract, meeting particular requirements of power, propulsion, and firefighting capability, although Jake finds plenty of other work for them when there are no LNG carriers moving.

The term *tractor* has not quite settled down in its meaning; there are people who are adamant still that tugs like *Freedom* and *Liberty* are not true tractors, as they have their propellers aft and don't actually pull themselves through the water. These boats have azimuthing drives, often known by the Rolls-Royce trade name *Z-drives*. Azimuthing drives can swivel around in any direction beneath the boat, much the way outboard motors can, and can be pointed in the same direction or in opposition to one another, giving the tugs tremendous power and maneuverability.

Most of the maritime world has accepted the term to mean any tug that is able to use its full strength in any direction, and that's what's required of the LNG escort tugs in Boston. (In fact, more and more harbors are requiring that more and more types of ships be escorted in, particularly tankers, and that work is almost always done by tractors.) "It's unnatural what these tugs can do," says Jake, who has been in the tug business for forty years. They can go frontward or backward with equal power, and they can crab sideways with speed and power, too, allowing them not only to give a shove to one side or the other, but also to stay perpendicular to a moving ship, ready to push or pull full-force if asked. Since the tug is able to move in any direction, everything she might be called on to do is easier: tethering up, docking, giving regular or emergency assist to a ship, or firefighting.

Firefighting capability is the main difference between the two Boston boats and earlier tractors built at the Washburn & Doughty yard in East Boothbay, Maine. In the mid-1990s, Bruce Washburn designed his first tractor tug, and since that time, the yard has developed a name for building tractors. At this writing, they have delivered seventeen and have

Left: *Controls on the port side of the helm on the tractor tug* Freedom.

Below: *The tractor tug* Liberty.

four more under construction. Says Jake of working with Washburn & Doughty, "That was such an experience! They're the finest people we've ever dealt with. The workers have a great work ethic and are very knowledgeable about their trades." Building vessels is not always so pleasant—disagreements often end up in court (some say always). Not with Washburn & Doughty. "In the two years it took to build those boats, there was not a cross word between the two companies. Anything we're ever going to build, if it'll fit in their shop, they're going to build it."

In the city of Boston, there is nothing on land or sea with the fire-fighting capabilities of *Liberty* and *Freedom*, says Jake. They've offered to help the Boston Fire Department if they ever can; there are connections on the boats to hook up to fire department equipment. "We can feed every fire engine they have." Five-hundred-horsepower diesels can pump 3,000 gallons of seawater a minute through each of the two monitors aboard each vessel. Additionally, each tug has a deluge system that protects her from the heat of a fire and allows her to get closer. Nozzles are spread around both deckhouse and pilothouse and spray water on the sides and windows.

"That deluge system is a sight to see—you almost can't see the tug. There's a shroud of water all around it. You have no worry about windows melting out."

The tugs carry powerful foam extinguishing systems, too, sized as the regulations require.

But with any luck at all, the tugs' firefighting powers won't need to be used other than for the dramatic demonstrations the crews enjoy putting on. "We can shoot water right up over the Mystic River Bridge," says Jake.

Jake's tugs are far from the only vessels accompanying an LNG tanker into Boston Harbor these days. By the time the ship arrives, the Coast Guard, local and state police, and the Massachusetts Environmental Patrol are also present. "The Coast Guard has a lot of little boats running around," says Jake, who mentions some forty-one-footers, one vessel maybe eighty or ninety feet long, and "some high-speed rubber boats with machine guns on their bows." People laugh at the rubber boats, asking, "What could they do?" What they can do is get somewhere in a hurry, and a machine gun staring you down does have a certain intimidation factor. The State Police have boats, the Boston Police Department has two, the Environmental Patrol has another, and over top of it all is a helicopter or sometimes two. A security "bubble" is kept clear two miles ahead of the ship, a mile behind, and five hundred yards on either side; nobody is allowed inside this area. They even close streets and wharves as the ship passes, and the Tobin Bridge is closed for about twenty minutes as a ship approaches and travels beneath it. At least sometimes, if not always, flights at nearby Logan Airport are halted as the ship passes, too.

The tugs turn the ship 180 degrees in the basin east of the Tobin

Bridge, and help her back up the Mystic River to her berth at the Distrigas facility in Everett, where she unloads. She's backed in so that if anything were to happen, she could quickly leave. On September 11, 2001, in fact, an LNG tanker was pumping off when the World Trade Center attacks took place. "In twenty minutes, we had the ship under way. We shot over, the lines were loose, and away they went."

The first time one of the two new tankers came in accompanied by both new tractor tugs, Jake walked out to the end of his dock to take a picture. Within moments, a police boat whizzed up, its guns at the ready.

"Who are you, and what are you doing?" the police boat crew asked brusquely. Jake told them that he was from Boston Towing, which owned the tugboats, and he wanted a photo to hang in the office. "Show us some ID," they demanded.

He produced a driver's license and then pointed to one of the tugs alongside the LNG tanker. "Look over there. That boat's named after me," he told them.

They were then very apologetic, but Jake said no, it was fine. In fact, he was glad to see how well the security worked. "But how did you spot me so quickly?" he asked.

Tugs Harold A. Reinauer II *and* Liberty *help the LNG carrier* Matthew *into her berth at Everett, with Coast Guard escort boat patrolling the area.* COURTESY OF THE U.S. COAST GUARD

"We didn't. He did," they replied, directing Jake's attention to a helicopter overhead. "Anything suspicious he sees on the ground, he reports to the security fleet surrounding the tanker, and we're dispatched to the exact spot."

Distrigas and the City

Every time a ship comes in, Distrigas makes a payment to the city of Everett that covers extra deployment of officers in the area. They've been generous supporters of new police department initiatives, too, says Chief Steven A. Mazzie. "In hard times, they're asking, 'What can we do for you?'" he says.

Mazzie has been with the Everett Police Department only a decade and a half, but he's third generation; Mazzies have been on the force since 1926. Though there are more precautions and more security measures taken now than before 9/11, to Chief Mazzie, Distrigas's facility is just part of everyday business. "Even the people that live down in that area are more concerned about crime than they are about that place." There's only a small residential area nearby, Mazzie says; most of the area is heavy industry. "When I was a little kid, I'd go visit Dad working security detail down there. He'd say, 'This is why your taxes are so low.' You don't realize how vital these companies are to the economy, what role they play. We have a close relationship with the company, and the other companies down there—it's a nice partnership with them."

From a policing point of view, Mazzie says his officers must be aware of people with a particular interest in the plant—suspicious individuals taking pictures or making drawings, for example. "We just check them out, make an inquiry about what's going on. It's no different than regular police work, just a little more TLC."

They're working on evacuation plans in case of an incident in Everett or in Boston. The plan includes a lighting system for the main arteries leading out of town. "We'll educate the people that when these lights are on, there's no parking, and two lanes will be open headed north, going out."

The ships have been coming and going for so long, it's routine. There haven't been many issues or problems with the company. "They have a good record. They want to have a good, safe business, and be good neighbors." The only problem Steven mentions is keeping the police officers fresh. "We gotta be careful we don't get lulled into a sense that everything's okay. But the guys are good."

The Northeast Gas Association
and the Massachusetts LNG Firefighting Academy

The Northeast Gas Association, formerly the New England Gas Association, is a regional trade organization representing all the aspects of the natural gas business in New England, New York, and New Jersey. Their focus includes education and training, research and development, and increasing public awareness of natural gas. Twenty years ago, when current Massachusetts state fire marshal Stephen Coan was head of the Massachusetts Firefighting Academy, the Academy and the New England Gas Association joined together to create an LNG firefighting school. "They provided technical assistance and monetary resources," says Coan. The Academy developed not only theoretical training, but also a flammable gas facility where students can observe and handle real-life situations.

Ten times a year, training sessions are held at the Massachusetts Firefighting Academy, a division of the Commonwealth's Department of Fire Services. The regular LNG course is a two-day program, which not only members of municipal fire departments but also utility company employees attend. "It's the first time those people ever see LNG," says Richard Swartz, a coordinator for the program. No one in an LNG facility ever wants to see the stuff; it means something's gone wrong.

The academy's program starts with a four-hour classroom session in which they describe the properties of LNG and give an understanding of how and why it reacts the way it does. Then they head out back, where they have LNG on-site and various props to help students learn how to handle the product. They have a large pit into which they can release LNG to see how it acts. They practice covering sewers so the liquid can't run down into the system, which would create a dangerous situation as it gasified and expanded in the enclosed area. "It becomes flammable, and if there's a cigarette down the road, you've got a problem," Richard explains. They show how the liquefied gas flows, and then ignite it so students can see how the fire slowly, gradually rolls back to the source. "It's very different from propane," Richard says. "Propane is very quick, very forceful."

First they put out the fire with dry chemical extinguishers, and then spray the pool to demonstrate why using water is not a good idea: It causes the natural gas to vaporize much faster than before.

As the pit is filling, the students practice how to control the vapors with hose lines. "We can disperse or direct it easily," says Richard. Again

*The Texas A&M University System Engineering Extension Service (TEEX)
regularly brings together chemical engineers, industry representatives, and
emergency responders to observe and extinguish controlled LNG fires at
its training campus in College Station, Texas.* COURTESY OF MARTIAL VOITIER/TEEX

they ignite the fire and then spray the flames with water, which makes
the fire more forceful and creates loud popping sounds like explosions.
"It's actually just ice forming and breaking away," Richard says. "That's
the highlight of the course, though, seeing the effect of adding water."

They also have an LNG tanker truck on site, and they demonstrate the
safety features built into that vehicle. They create a leak and practice cutting off the fuel. "With flammable gas, any time you have a fire and can't
shut it off, you want to leave it burning. If you get it out, the remaining
gas is only going to be looking for another ignition source." It could be a
spark, a flame, or just something hot enough to set it off again, and by
then, the gas might have migrated to a more treacherous location.

Periodically, the Academy also holds a Flammable Gases Incident Command Training, which has been attended by managers of large city fire
departments from all over the country. Commanding the war against an
incident is a critical role (as was demonstrated negatively during Hurricane Katrina in 2005). The person in that position must arrange for a network of technical advisors and know how to deal with them once they're
on the scene.

The members of the Northeast Gas Association have also created a
Voluntary Assistance Network; any one of them will respond to any gas-
related emergencies in their area. Local gas companies respond with

specialized equipment and provide technical assistance to the local emergency crews.

In the twenty years or more that the program has been in operation, thousands of firefighters and industry employees from Massachusetts, neighboring states, and beyond have participated in a unique training program combining classroom principles and real-life scenarios involving LNG, propane, and natural gas. "The benefit is we have an educated industry and fire service in our area," says Fire Marshal Coan.

Trucking and LNG Peak-Shaving

The interesting thing is that LNG is so much a part of our everyday life already. About two hundred LNG facilities are in place around the world, including more than a hundred in the United States in addition to the import terminals. Within California in 2004, nine liquefaction plants were running and others served the state from Wyoming, the Pacific Northwest, and Arizona. These produce fuel for transit authorities, trash haulers, and other heavy-duty trucks, including some serving large grocery chains. Twenty-eight private or publicly owned LNG vehicle-refueling

The control panel at the Distrigas truck loading area.
COURTESY OF DISTRIGAS

stations are in various parts of the state. Some of the tanker trucks are equipped with vaporizers that allow LNG to be trucked to a site in need of natural gas for immediate use.[23]

The concentration of LNG facilities is in the Northeast, however. Natural gas has always been sold at a premium in that area due to its distance from energy resources. There is also a higher dependence on natural gas for electricity than in other parts of the country (40 percent, as opposed to 20 percent nationwide). The LNG facilities in the region are mostly peak-shavers and satellite plants, which store liquefied natural gas to draw upon during times of increased electrical demand.

Because it takes up one six-hundredth as much physical space when liquefied, natural gas is considerably easier to store in that form. This is particularly helpful in bony New England, which has no underground storage spaces such as are utilized in much of the rest of the country. The peak-shavers—fifty-seven facilities in New England at this time—receive gas by pipeline and liquefy it when economic conditions are right. New England's thirty-nine satellite facilities have no such ability. They receive the product by truck, either from import terminals such as Distrigas in Everett, where about 10 percent of the LNG imported goes out by truck, or from a plant in Hopkinton, Massachusetts, that takes pipeline gas and liquefies it in part specifically to sell via truck.

Even the plants with liquefaction capabilities sometimes bring in LNG by truck—it all depends on the dollar numbers. Sometimes it's cheaper to pay for trucking than to liquefy, and sometimes it isn't. Peak-shavers with liquefaction capability sometimes find their capabilities to be a good negotiating tool with the pipeline people; if the utility doesn't like the price offered, it can just truck LNG in from somewhere else that might offer them a better deal. (Hereafter, we'll take the liberty of lumping peak-shavers and satellites together and referring to them all as peak-shavers because they both have the same storage function.)

The legal aspects of the contracts between gas suppliers and utilities are complicated and varied, but are a big part of the overall picture. On a simplistic level, utilities know that there will be times when demand for electricity—and therefore natural gas—will be higher than the amount they have contracted for. In winter, these periods might come only on the coldest days and last only for a few hours, say between six and nine in the morning and again in late afternoon and early evening, as people take showers and start the laundry and use their electric stoves and space heaters. In summer, peaks might come during the heat of the day, when air conditioners are running hardest. During hours of high electricity use, the utility can regasify and burn the fuel it has stockpiled, which is often more economical than buying gas over and above their contracted amount, as the price invariably rises at times of increased demand. There are even times when the pipeline is unable to deliver all that a utility needs. A peak-shaver allows the utility to manage peak loads without its customers even being aware that there might have been a shortage.

"On any given day, there are a hundred LNG trucks on the road in the Northeast," says Chris Hosford, whose company, CHI Engineering Services, Inc., specializes in the technology of natural gas, LNG, and other cryogenic gases. Carrying ten to twelve thousand gallons of LNG apiece, the trucks deliver the product to peak-shaving facilities around the region. One peak-shaver alone, in Providence, Rhode Island, gets 2,000 truckloads a year from the Distrigas facility in Everett, Massachusetts. (KeySpan applied to expand that plant so as to import LNG from abroad by ship, but in 2005, FERC turned them down, saying that the storage tank and its environs, constructed long ago, were not up to current standards, and noting that the required protective exclusion zones for the facility would extend offsite.)

New peak-shavers are built from time to time; at this writing, one is under construction in Waterbury, Connecticut, and another was finished a few years ago in the Berkshires. Most of these, doing their business within their home state, are regulated by the states, which have hoops to jump through very similar to the federal ones required of new LNG terminals or other facilities. Most states have simply adopted the standards, rules, and regulations of the Federal Energy Regulatory Commission, so the engineering, construction, and operation of such facilities, at least the more recent ones, are the same countrywide.

LNG tanker trucks are also used to bypass natural gas pipelines temporarily out of service, perhaps for maintenance.

There are a few niche market users of LNG as well, as California has well demonstrated. One goal is to replace propane in some industrial sites. Because propane is delivered under pressure, the tank must be considerably heavier. Given the same weight and length, a trailer of propane will contain about only half the energy that an LNG trailer will, making LNG less expensive to deliver at greater distances on the road. In sparsely inhabited areas of the country distant from the pipeline grid, LNG is being used for mining and forest product facilities.

The other niche, one that some see as having great potential but others dismiss entirely, is for vehicular fuel. At times, LNG has been less costly than gasoline or diesel, and historically, it burned more cleanly than either. The new ultra-low-sulfur diesel eliminates that advantage.

Compressed natural gas (CNG) can also fuel vehicles, and is often used in areas with access to natural gas pipelines and particular concern about air quality, like California, or for strictly economic reasons. A number of municipalities use natural gas in some of their vehicles—can more than 4,500 California buses be wrong? (Rhetorical question.) In Texas, the largest fleets of vehicles making the switch were bus companies; but one, the Houston METRO bus system, one of the earliest adopters of LNG, has switched back to diesel, albeit the new, cleaner diesel systems. "The policies were a bit ahead of the technology," says Dub Taylor, the director of the Texas Energy Office. Fueling stations were more like prototypes, not yet commercial units, and there were operational problems. Faced with a fixed budget and routes and schedules to keep, they felt they had to return to the tried-and-true diesel.

With the help of federal grants, some in private industry started making the switch to LNG a few years ago: in 2001, the Sysco Foods distribution center of Houston committed to a 90 percent shift to vehicular use of LNG, and has followed through. "Absent public funding, I doubt if they would have," says Taylor. While there was interest in LNG, he says, it peaked a few years ago and has been on the decline as natural gas prices have risen. "For LNG, now, there have to be drivers other than financial." Some heavy-duty vehicles that travel a lot of miles in areas such as in California, with its rigorous air quality requirements, can't carry enough compressed natural gas, so they have an incentive to use LNG instead, says Taylor. "There has to be enough throughput at a fueling facility—it's hard to make it pencil out."

At Los Angeles World Airport, more than 300 vehicles run on natural gas, either compressed (80 percent) or liquefied (20 percent). The Port of Los Angeles has introduced its first LNG-fueled yard tractors and is actively seeking out other options for reaching the required low emissions standards, including LNG-powered equipment.

This is, however, an effort of decreasing importance as far as air quality is concerned. With the ultra-low-sulfur diesel fuel headed for the pumps in the lower forty-eight in 2006, and the coming of vehicles that can utilize its characteristics to the fullest, LNG will offer no advantage where air quality is concerned.

As of 2005, because of its abundant supplies of natural gas at low prices, Argentina had by far the most natural gas vehicles. Pakistan had a large number as well; at the end of 2005, there were more than 650,000 CNG-powered vehicles on the roads there, with 55,000 more being converted each year.[24] The government of Punjab mandated that all public transport vehicles run on natural gas by 2007.

There are disadvantages to compressed natural gas produced from pipeline gas, in addition to the fact that because it's under pressure, it requires heavy tanks that for their size are limited in what they can hold. A serious problem is that the source gas varies considerably as it comes out of the pipe, containing varying amounts of water, hydrogen sulfide, oil, particulate and foreign matter, and propane; sometimes gas suitable for vehicular use is not available in the pipeline at all. Using LNG as the feedstock for a compressed natural gas fuel minimizes quality issues, as it has been processed prior to liquefaction and is nearly pure methane.

Vehicles also run on propane, which, like the other natural gas products, is clean-burning and suitable for use indoors. Its availability in small canisters is convenient for the owners of a few forklifts and the like who can simply fill or swap out propane bottles. It would be cost-prohibitive to invest in the infrastructure needed for either LNG or CNG for such a small fleet.

In the United States at this time, the economic advantage of natural gas for vehicular use has largely disappeared, however. "The uncertainty of the natural gas market has people leery," says Dub Taylor. "While pricing of petroleum is equally volatile, the challenge with natural gas now is that the demand from non-vehicular uses—electricity, specifically—has sustained new floor prices." Without developing new domestic sources, or importing more, the prices will stay.

Another aspect of the issue is that regardless of the national output of petroleum, or the prices at which it's available, petroleum can always be imported. "You can import finished gasoline," says Dub. "But we're more limited in our ability to import natural gas."

A Case Study in LNG Fleets: Dallas Area Rapid Transit

The Dallas Area Rapid Transit (DART) has in its purview thirteen cities and a rail connection to Fort Worth shared with that city's transportation system. DART operates 130 bus routes, forty-five miles of light rail, which is planned to more than double in the next ten years, thirty-one miles of freeway marked for high-occupancy vehicles only, and other services.

At this writing, DART operates 691 buses, including 184 that run on LNG. Since 2002, the diesel buses have been running on the new "clean diesel" fuel. Studies DART and the state ran showed that emissions of nitrogen oxides were down a third from old diesel, and the particulate level was so low they had difficulties in testing it: sometimes what came out of the tailpipe had fewer particulates than the ambient air![25]

The LNG fleet was introduced in 1998 with the help of a federal program that was addressing clean air issues, first with just ten buses, but soon coming up to the full number. Mike Hubbell, DART's vice president of maintenance, chose to work with LNG in part because of his belief that ultimately it might be a cryogenic fuel—hydrogen, he assumes—that would eventually prove to be "the fuel of preference." He wanted to get over the hurdles that cryogenic materials raise.

Dallas doesn't lie along major gas transmission lines and can't tap the fuel and liquefy it locally, so LNG comes to each of Dallas's two fleet fueling stations by truck from one of three liquefaction sources, in Texas, Alabama, and Wyoming.

Mike's goal was to make the fleets using different fuels integrate smoothly. The fueling facilities cost about four million dollars, and footprint space was set aside at each for future hydrogen extractors. Some of the design issues have nothing to do with the particular fuel, but rather the surrounding area. The LNG tanks are double-walled, stainless steel, vacuum-wrapped, but a fire marshal made them lay their LNG tanks down lengthwise even though it would be a more economical use of space to stand them up. "He was afraid of gunfire from the freeway," says Mike.

The stations have been designed so that any bus can come to any island at the station and receive its fuel. "It's not quite as simple as regular and premium automobile gasoline," he says, but every servicer can fuel either variety of bus. "We didn't want to treat the LNG buses as bastard children," he says. At one station, they fill 197 buses in a six-hour period. Nobody's afraid of the stuff—everybody knows what precautions to take, just as they know what precautions need to be taken with diesel. The filling attachments latch into place and seal before the valve opens to release the fuel into the bus. The attendants wear gloves and goggles, because on very rare occasions when the seal gets worn, there might be a little leakage when they remove the apparatus. "It's so cold, it's hot!" says Matthew Gardner, who fuels up fifteen LNG buses and fifteen diesels every night. "A few times, you get in a hurry," he says, "and you might get a drip. You definitely feel it. It's not just like 'Oops'—it blisters up—but you're over it in four days. It's like you burned yourself." The only disadvantage to LNG that Matthew reports is that it takes more fuel and more time to fill a bus than with diesel. "I'm pretty used to it, though."

"We keep pretty safe," says Rodney Kent, Matthew's supervisor. "We know if it were to spurt out, it'd be dangerous, but it almost never does. Cold is the most dangerous part about it. If it leaks, it evaporates so quick it would put a fire out."

"It gets pretty simple," says Matthew. "Almost boring."

There have been no significant injuries with LNG at DART. "Ultimately, we've made LNG as reliable as any other fuel," Mike says. He does add, "Not without a lot of effort." They were among the first users of the technology, and their learning curve has benefited subsequent transit companies interested in LNG.

As they are retired, DART's old diesel buses are presently being replaced with new (clean) diesels. No LNG vehicles are due for replacement until 2010, which is a year when emission standards are to change radically once again. It's too soon to tell how the new buses at that time will be powered. "It is now an economic issue," says Mike. The first year, the ten original buses were more expensive to run per mile than diesels, largely due to a number of problems that were worked out during that time. The second year the operating cost per mile for the LNG vehicles was about 8 percent less than the diesels' even though fuel cost was higher—diesel buses average about four miles to the gallon, but those that run on LNG get less than two. The operating cost is now higher again. "Assuming LNG's price to be constant where it is today," Mike says, "the breakeven for diesel is $2.45 a gallon, and I'm now paying $2.10."

Another factor is that at this time LNG buses cost $350,000 each compared to $300,000 for diesels. (Hybrid gasoline/electric vehicles, presently the ones that show least emissions, cost another hundred thousand apiece, and hydrogen fuel cell–powered buses, undoubtedly the cleanest, cost $900,000, twice as much as hybrids. Mike believes that, ultimately, hydrogen fuel cells will be the answer, but he says that technology isn't likely before 2015. A lot of advances have been made—the engines are smaller than they used to be, they create less heat, and the price has already come down radically—they used to be $1.5 million. Still, the technology is far from ready for commercial deployment.)

Following a random conversation once with the director of the state general land office, Mike had an unusual opportunity, puzzling out and discussing and negotiating a most intriguing LNG project. It was his idea that DART might join with the state on a project to liquefy gas on the site of one of the 22,000 economically stranded gas wells on the millions of acres of land the state holds mineral rights to. (Stranded gas is that which is so far from a distribution system or so impure that it isn't economically viable to harvest.) But if DART were to purchase a liquefaction unit for a couple of million bucks, then the whole economic formula would shift for both sides. The state would get to sell gas that it hadn't been able to, and, in return for installing the liquefaction unit, DART would get fuel. That the source gas might be impure isn't a problem with liquefaction, as all the unwanted matter drops out along the way. (And there's a market for it. Water, of course, is nothing, but propane, butane, and the other heavies are saleable.)

The board of directors wondered if it would be inappropriate for a governmental agency like DART to be in an entrepreneurial-risk situation, even though nearly all of them are personally. "Isn't this wildcatting?" they asked. But no, Mike said. "The state of Texas says it has this gas in the ground, and I can get it out—that's as low a risk as you can get. It's not as if we want to get rich from it—you know the old saying, pigs get fat, hogs get slaughtered."

Mike says he would have had only a three-year payback time if DART had liquefaction capability and free or even inexpensive gas to feed into it. "Keep a bus twelve years, seven hundred buses with fuel costing fifty cents instead of $2.35—that's a lot of money saved!" Another advantage would have been that with gas markets so volatile, the cost of fuel would have stabilized, making it far easier to project out and budget for a number of years. As the operator of a fifty-six-million-kilowatt-hour rail system, Mike is very pleased that the state entered a six-year agreement for electricity at a fixed rate, 8.56¢/kwh. "I'm paying eleven cents at home—that saved us millions of dollars this year," he says. If diesel goes up a penny, it costs DART $67,000 a year. The recent roller-coastering of gas and oil prices makes it extremely difficult to plan ahead, and this LNG project might have made a lot of sense.

But it was a Catch-22 situation. If it were economically viable for someone to tap those wells, by definition, they would no longer be stranded, and they would not be available to DART. And sure enough, that was the trap that killed the idea. Mike says he hasn't given up, though.

An LNG Success Story: Cheniere Energy

In the mid-1990s, Cheniere Energy had determined that exploration for natural gas in the Gulf of Mexico was not going to be as profitable a venture as they had hoped. The costs of drilling were going up while the reservoirs were getting smaller, all across the Gulf. It's a sign of a maturing resource base. "People pick the low-hanging fruit first," goes the saying. But the people at Cheniere were certain that there was an expanding market for natural gas in this country, and that, somehow, there was a role for the company in that future. After considering various exploration options, they started thinking about LNG. The convention of the time was that LNG needed to bring $5 to $6 per thousand cubic feet to break even, but they found that it could be landed for $2 to $3.50, depending on where it came from and the cost of transportation. With domestic exploration and development costs approaching the same price

Artist's rendering of Cheniere's Sabine Pass LNG receiving terminal. At the shore, a membrane-tank carrier is shown on the left and a Kvaerner-Moss design tanker on the right. Each of the six storage tanks is surrounded by a berm. On the upper right are ambient air vaporizers, where the LNG is regasified. More vaporizers sit between the ships and the storage tanks. The four units on the right are the terminal's electrical generators.
COURTESY OF CHENIERE ENERGY

range, it looked like a competitive solution to the country's need for new sources of natural gas supply.

So then the question was, where in the LNG value chain could a small company participate? Upstream was not an early option—big balance sheets and strength are necessary to talk with governments overseas. The majors had that all wrapped up. Liquefaction was capital-intensive, requiring upwards of $5 billion per project. Shipping wasn't attractive for a young company at that time either. But domestic receiving terminals, on the other hand, seemed like a good niche.

There'd been no new terminal since the 1970s, so the question was where to locate. Cheniere's cofounder and chairman Charif Souki listed the necessities: a shipping channel deep enough and with good port facilities and yet not too congested to accommodate very large LNG carriers, a large parcel of land away from residential areas in order to maintain all the required safety zones on the site, and reasonable access to pipeline infrastructure. Finally, and most importantly, they needed community support. The chairman of FERC at the time, Pat Wood III, had pronounced to all industry hopefuls that without local support, they might as well pack their bags because there was nothing FERC could do.

The Cheniere team looked at every channel from Canada to Mexico, and landed back at the Gulf of Mexico. That's where there's the most consumption of natural gas, as well as the pipeline network to move it elsewhere. The ship channels exist, the land is there, but although certainly Texas and Louisiana are accustomed to oil and gas business, the company still would have to earn community support before pursuing a project.

From the start, the new company chose to start at the bottom, holding community forums and actively seeking out opposition, so as to address concerns from the beginning. Although there was skepticism at first, as Cheniere worked they earned local trust. Instead of making pronouncements the way some of the major oil companies are apt to do, they worked to garner support first with the community, which then spread and made it easier for local political representatives to support them, then mayors and congressmen, senators and governors.

The first project was in Freeport, Texas, and Cheniere was among the first companies in the continental United States to make their way through FERC and successfully certificate a project. In fact, LNG was facing so much opposition on the East and West Coasts, FERC almost stopped the review process on all projects, but Cheniere got support not just from

Gulf Coast stakeholders, but from the upper Midwest, too. It wasn't simply a matter of jobs from the LNG terminal for the Gulf Coast, but of access to low-cost natural gas, which industries and homes throughout the central part of the country depend on.

Early on, the company found it hard to raise the money for their first project. Wall Street didn't foresee the coming call for natural gas, and didn't think that the small young company would be able to get the permit. In 2002, the chief executive officer personally lent the company $20,000 to meet payroll for their thirteen employees. They sold 70 percent of the project, keeping a 30 percent limited partnership interest, and the partnership marketed the full capacity of the project for twenty years to ConocoPhillips and Dow Chemical. There was no longer any question about their capabilities.

Construction started on the Freeport facility in January 2005, by which time Cheniere was at work on three more projects in Texas and Louisiana. The company started the process to permit its second project, the Sabine Pass LNG receiving terminal, in 2003, raising $14 million along the way, and received the authorization to construct early in 2005. For that project, too, they signed up large tenants, Total and Chevron, for twenty years' regasification capacity, bringing in a guaranteed $250 million revenue each year, and shortly after the facility had broken ground, they'd already made application to FERC for authorization to expand the terminal. They raised $2 billion to execute the project, between a public stock offering, financing from a consortium of forty-seven banks, notes, and a term loan.

Cheniere's third project, in Corpus Christi, was a challenge with a very happy ending. They presented four possible locations to the local community and found opposition to them all from one group or another, including the local Sierra Club chapter, but Cheniere kept trying. The fifth try turned out to be ideal both for the company and for the community. It was an industrial site, near pipelines and far from residential areas, but more importantly for the environment, they would be able to return an industrially scarred area of six hundred acres to its natural state. Not only did the Sierra Club elect not to oppose the site, they actively supported the project before FERC.

Most recently, Cheniere started the permitting process for its fourth project, at the mouth of the Calcasieu Channel in Cameron Parish, Louisiana. They've expanded their interests in a number of LNG-related areas, too, including pipelines to connect their terminals to the existing interstate and intrastate systems, and an interest in an LNG shipping venture.

Cheniere now has more than 130 employees, and there's no question about delivering their paychecks. Shareholder returns have also been remarkable: through 2005, the company's stock was proclaimed to be the best performer during the previous three years and the previous five years in the universe of companies surveyed by the *Wall Street Journal.* "It all grew out of being humble and open and transparent in [our dealings with] communities," says Cheniere's corporate spokesman David Castaneda. "It's a wonderful company to be a part of."

Deepwater Operations

OFFSHORE facilities have been in use in the oil and gas business for more than thirty years; it is only logical that the concept would come to the LNG industry. Three general technologies are proposed. The first uses a ship with both storage and vaporization capabilities. A number of different companies are proposing this kind of technology, and one, Excelerate Energy, is today producing natural gas at their Gulf Gateway. The second and third approaches have fixed regasification units, either floating and moored or on platforms, with differing schemes for storage.

There are advantages to locating an LNG facility offshore. The most obvious and perhaps important is that it can be placed near to the end market for natural gas without putting a populated area at any risk, however slim. There is no addition to port congestion—estimates show that the volume of imports passing through our ports will double in the next twenty years or so, and any ship that can off-load its cargo offshore is one fewer that will be contributing to harbor traffic. In addition, emissions from the ship or processing are removed from inhabited areas. Larger vessels can be accommodated than in existing onshore facilities. Placed well, offshore facilities can utilize existing pipelines with only relatively short connections required. Real estate, often pricey on the shore near urban areas, is also not an issue.

There are, of course, disadvantages. In some cases, the projects cost more to build than onshore facilities, and there is opposition to offshore terminals, just as there is to any LNG facility, no matter where the location. Some regasifiers use seawater as their heat source, and there is concern about the impact this—and the presence of the facilities and the ship traffic—may have on sea life and fisheries. Other issues are raised in specific locations. Scale and visual impact are a concern on Long Island Sound and in the Santa Barbara Channel, but not, apparently, on the Gulf of Mexico off Louisiana; perhaps that area already has been entirely

71

written off in that regard. "We're very careful to say we don't have one hundred percent endorsement," says one offshore developer, "but even the greens recognize there's an energy need, and natural gas is very good for the short term. Comparing dredging wetlands with developing off-shore, offshore is disliked less."

Practically speaking, while the technology is proven in the Gulf, some mariners question its applicability in other waters, specifically the North-east. Master Mariner John O'Reilly, who has years of experience in both the Gulf and the Atlantic, says that whatever the wind direction, the rel-atively shallow and entirely unprotected Gulf develops a short, steep sea, very uncomfortable for tugboats and smaller vessels, but not as much trouble for big ships like LNG carriers. In an easterly off Massachusetts, however, a long swell develops that might be problematic for larger ships. Passamaquoddy Bay pilot Bob Peacock worries about the tempera-tures, too. "Never saw freezing weather in the Gulf," he says. The cold is not just unpleasant, but leads to ice buildup from freezing spray. "I know the Gulf can get a lot of hurricanes, but a true northeaster is something to behold at sea."

On the other hand, the technology was originally developed in the North Sea, which has sea conditions about as horrendous as can be found anywhere.

Bob Arvedlund, retired chief of FERC's Environmental Review and Com-pliance Branch, thinks the offshore import terminals are a novelty: "I'm not saying it won't work. I'm one of them engineers that want to see my project. I want to drive up and see it. Maybe that's just me."

As of September 2006, in the United States there were fourteen pro-posals for new onshore facilities and expansions, and two accepted and eight new offshore proposals. The offshore scheme may be a relatively new idea for LNG terminals, but it has appeal and has been demonstrated to be effective at least in one location.

Excelerate Energy

The first American LNG import terminal to open its doors since 1982, Excelerate Energy's Gulf Gateway doesn't actually have doors. Excelerate is a new company with a new idea—new at least to LNG; in fact, their Energy Bridge system uses technologies well proven in other areas. Theirs is an all-in-one offshore regasification and delivery system that utilizes specially designed ships with their own built-in regasification capability that are able to mate up with an undersea pipeline connection. The vessels can also unload natural gas into a conventional pipeline, or lique-fied natural gas at a conventional terminal, but their unique ability lies in providing natural gas to a country without landing on its shores.

Not exclusively in the transportation business, Excelerate also delivers gas downstream. In fact, they see themselves as having a role to play sim-ply regasifying from other ships that have brought LNG from overseas and transferred it to their vessel, either at sea or across a dock, and they made their first at-sea ship-to-ship transfer in September of 2006. They could also bring natural gas to a temporary market somewhere, should the need arise. One of their major advantages over shoreside facilities is that they can be up and running about a year from a favorable record of decision from the Maritime Administration.

Despite its newness in the LNG field, the Energy Bridge system is made up of well-proven components. While the ships' design is state-of-the-art, it is also ordinary, like the other vessels coming out of the same yard, only with the addition of more equipment, and the vaporization units are the same as would be found in a shoreside facility. The buoy system, with its flexible riser through which natural gas passes into the pipeline, has been used for several years in the petroleum fields of the North Sea carrying crude oil the other direction, from wellhead to ship.

The Energy Bridge operation is largely automated, but can be man-aged manually. The ships have Differential Global Positioning Systems (DGPS) aboard, which give ten-meter accuracy. As a ship approaches, the submerged buoy sends out a signal, "Here I am." When the ship picks up that signal, her automated approach system puts her directly over the pipeline buoy, which has a messenger line floating on the surface, marked with lights. From the ship, that line is picked up and led through the watertight compartment in the bottom of the ship. The line is winched in, hauling the buoy up and into its place in the compartment, where it's seated and held by hydraulic rams. The compartment is pumped dry so people can work inside it, and a swiveling high-pressure fitting connects the ship's vaporization equipment to the export pipeline, through the buoy and flexible riser. (Animated video clips of the system can be seen on Excelerate's Web site: www.ExcelerateEnergy.com.)

The subsea component of the Energy Bridge system comprises the buoy itself with its swivel and moorings. Known by the trade name Sub-merged Turret Loading (STL) buoy, it is conical in shape, thirty-five feet high, twenty feet in diameter. It reaches neutral buoyancy at ninety feet below the surface, so when not connected to a ship, it rests well below

Excelerate Energy's Energy Bridge system, showing the buoy hauled up into place in the ship, the buoy anchor lines, and the flexible riser through which the vaporized gas flows into the pipeline in place on the ocean floor. ARTIST'S RENDERING COURTESY OF EXCELERATE ENERGY

the draft of any vessel coming along. (Even so, the company anticipates having a safety zone with a diameter of one kilometer and a 2,000-meter diameter no-anchoring area around each of the two buoys they're planning at their Northeast Gateway site.) The buoy is moored in place by eight anchor lines to the sea bed, and a flexible pipeline runs through it to connect the ship to the pipeline on the ocean bottom. The buoy and its anchoring system serve as the mooring for the ship, as well, and the ship can freely rotate around with the wind, just as if she were on an ordinary anchor. The system has been designed to withstand weather and sea conditions to the level of a hundred-year storm, and, in fact, rode out the 2005 hurricanes in the Gulf with little difficulty.

The center of Hurricane Katrina passed within 150 miles of the Energy Bridge system. "It was rough out there," admits Rob Bryngleson, vice president of development and downstream services. "The ship went through a ten- to fifteen-degree roll, but we kept producing with no problems. Pipelines called, saying, 'Hey, can you guys give us more?' We were up to 700 Mcf a day, but then some pipelines downstream were damaged and we had to cut production down. We could give them everything they wanted and never had any stoppage."

The Gulf of Mexico is riddled with pipelines; the particular location of Excelerate's Gulf Gateway deepwater port was chosen because of the particular pipelines near that spot, 116 miles off the coast of Louisiana. They have connections there to two lines—the Blue Water and the Sea Robin pipelines—which, in turn, come ashore to the major transmission lines near Henry Hub, the trading point for the New York Mercantile Exchange natural gas contracts.

In areas where nearby pipelines aren't already waiting for their product, Excelerate would choose to locate about twelve miles offshore, as their Northeast Gateway Energy Bridge deepwater port off Massachusetts is planned to be. "It's just off the visual horizon," explains Ed Scott, director of development at Excelerate. The company has no wish to upset anyone who felt the Energy Bridge would clutter up their view. The amount of clutter provided by a single ship, even a big one, only present some of the time might not seem excessive to everyone, but some people have been opposed to offshore facilities on just those grounds. "Our concession to that component of the opposition is to put it twelve miles out," Ed says. The distance should be reassuring to people concerned about an incident at the facility, too. Should anything go wrong, the only members of the general public possibly at risk would be the few fishermen who might be in the vicinity.

Excelerate likes 250 feet of water or more. (They have model-tested the system to a little more than 3,000 feet, and theoretically it could work as deep as 7,000 feet.) In Massachusetts Bay, few choices were available for placement of a facility (and, in fact, the other company with a similar idea, Suez, ended up with their Neptune's proposed spot only about five miles from the Northeast Gateway's). In addition to the shipping lanes approaching Boston, there are shoals, one federal and two state marine sanctuaries, and a state disposal site where dredge spoils are dumped. "You overlay all that onto a map, and there's not much left," says Rob. "Only a little triangle." The water depth in the triangle is 270 to 290 feet, barely exceeding the minimum requirement, but Massachusetts Bay was a good location nonetheless because it's close to the needy urban market of Greater Boston. The Hubline gas pipeline already crosses the bay from Beverly to Weymouth. Going offshore also means they do not have to acquire a vastly expensive bit of shoreside real estate.

An offshore system like Excelerate's is relatively inexpensive to pull together, Ed says, though one of the costs that is higher at sea than on land is the pipeline to connect into the system. Pipeline is not trivial, with a total cost of a million or two per mile on land and even more subsea, depending on the topography and the time of year. But in other respects, the cost of building offshore is considerably lower than on land. The actual installation period runs only about eight months, compared to three years on land. With three ships already operating and another on order, with varying arrangements as to leasing or ownership, the company has a good start on the rest of the project.

It is Excelerate's commitment, Ed explains, to make each project as environmentally friendly as possible. Two of the issues that get raised during the application process are emissions from the ships and water discharge from the processing. Gulf Gateway does have an open-loop regasification system: it brings in seawater, runs it through a heat exchanger to warm and gasify the LNG, and discharges it again. There are concerns that this system may have a deleterious effect on the marine life in the surrounds, so the Energy Bridge vessels are designed to be able to switch to a closed-loop system. Excelerate has committed to using the closed-loop system at the Northeast Gateway. "In open loop, the water chills thirteen degrees, worst case, from entry to exit," Ed explains. "The plume doesn't extend more than a hundred meters, not even the length of the ship away. They're monitoring it closely, but the jury's still out on open loop." Rob says that the seawater of Massachusetts Bay is only marginally warm enough to use in the winter, anyway.

Looking along the deck toward the bow of the Excelsior, *one of only three tankers in the world (all owned by Excelerate Energy) specially equipped to off-load LNG to a deepwater regasification facility. For scale, notice the man just left of center foreground.* Courtesy of Excelerate Energy

There are downsides to a closed-loop system, as well. It uses more fuel, which makes for a higher cost of processing, and the ship's boilers have to be bigger than they otherwise would be. Also, more fuel burned means higher emissions from the vessels, but Excelerate has made design changes that will reduce air emissions so much that they will no longer be subject to the EPA's new point-source permitting requirements. The company has also developed an improved system that uses the heat of the ship to warm the vaporizing loop and the chill on the other side of the loop to cool the ship's generators, and even within the closed-loop system has reduced water usage from fifty-four million gallons a day to just under three million. "It's expensive, but it's the right thing to do," says Ed.

It is a slower process, however. An open-loop system can handle 690 million cubic feet per day, whereas a closed-loop system can only take 400 or 450 Mcf/d. Having two buoys at the Northeast Gateway would double the output potential.

Summing up, Ed says, "We've had the good fortune to be on the cutting edge of deepwater development, but there's a bad side to that, too. The regulatory process has taken a longer time than we expected."

Excelerate is looking at other sites—they've announced interest in California and in Florida, and have a couple of other areas in mind in U.S. waters. They're expanding abroad, too—in early 2006, they announced a partnership to develop an offshore system in Pakistan. Although there are a lot of LNG import projects under discussion countrywide and some have filed for final licenses, Ed believes there are ample economic reasons for Excelerate to continue their search. "We make a lot of sense," he says. "We're not the biggest, but we don't take a lot of room and we're not ecologically intrusive. We're infinitely flexible."

Other Offshore Plans

Two other offshore facilities, projects that have been developed by major oil companies, have already been approved in this country, as well as one off Baja California. They are gravity-based terminals with storage in concrete structures that rest on the ocean floor and regasification on-site. None of these is presently being built, however; their system appears to have proven too expensive or difficult to build.

Other styles of offshore terminals are also under consideration; these facilities use a fixed construction or a moored vessel for the regasification systems and storage. One is proposed for Long Island Sound. Broadwater Energy, a partnership between TransCanada Corporation and Shell Oil, wants to build a facility that would appear to be simply a large ship moored in the sound, nine miles off Riverhead, New York, and eleven miles from the Connecticut shore. The vessel, a Floating Storage Regasification Unit (FSRU), would be somewhat larger than the *Queen Mary II*—about 1,200 feet long compared to her 1,132 feet. The FSRU is also wider, at 180 feet; the *Queen Mary II*'s beam is 147 feet. The FSRU's deck would stand 75 to 100 feet over the water, which, although tall, is much lower than the *Queen Mary II*, where passengers on the upper decks stand eye to eye with the 151-foot-tall Statue of Liberty. It's a big box, certainly, but from the shore, on either side, it would be barely visible on the far horizon.

The particular site was chosen after considerable research into the alternatives for both shoreside as well as offshore. The sound was appealing to Broadwater because it offered protected waters as well as its proximity to the greater New York City market, yet it avoided dense population centers. After considering other uses of the area—shipping lanes, ferry routes, fishing, and recreational usages—they chose a spot on the broadest part of the sound, in seventy to ninety feet of water.

An LNG carrier would come in after her voyage from overseas, and, assisted by tugboats, tie on. She would off-load to tanks in the FSRU just as she would at a facility ashore, and be on her way. The FSRU would store the LNG until it was regasified, and then the natural gas would go ashore to the existing Iroquois pipeline via a twenty-five-mile underwater connecting pipeline. As of 2005, the company forecast that their first delivery would be in 2010, but the project is facing serious opposition from many sides.

BHP Billiton has filed an application for a similar import facility in the Santa Barbara Channel, while in that same general area, with their Clearwater project, Crystal Energy has applied to reconfigure an old, inactive oil platform a little more than twelve miles off Ventura County. Clearwater would install transfer and vaporizing systems on the platform. LNG carriers would tie up to a floating dock alongside the platform just as they would shoreside, and a new pipeline would take the regasified product to an existing Southern California Gas Company pipeline near Camarillo. A third project off the southern California coast was by an Australian company in the summer of 2006, similar in type to Excelerate's.

Another interesting concept just crossed the horizon: gas by wire. Instead of regasifying and piping the natural gas ashore, there's talk of

combining a floating receiving terminal and a power station, thus generating electricity offshore and sending just the power ashore.

Placing LNG import terminals ten miles or more offshore has a lot to
recommend it. None of the serious scientists today are predicting that
even the worst LNG-caused fire would affect an area as much as ten miles
across. The most serious complaint about offshore regasification units
seems to be that open-loop systems may have a deleterious impact on sea
life. Closed-loop systems don't attract that criticism.

LNG Carriers and Their Crews

THE whole reason for liquefying natural gas is to move the gas around, so the ships are an important factor. At this writing, we are closing in on 200 LNG carriers worldwide, flagged from twenty-two countries; the majority are from Europe, Japan, Malaysia, Panama, Liberia, and Bermuda. (There are no longer any American ships. More about this later in the chapter.) As of 2006, more than a hundred new carriers are in the works at the nine yards that build LNG ships, in South Korea, Japan, Europe, and now China. The price tag has come down from a reported high in the mid-1990s of $280 million.[26] Ships that can carry the typical 145,000 cubic meters of product are cheap now, somewhere between $150 to $160 million, the same figure that bought an American-built ship not that much smaller in the 1970s.

As are the ships carrying every other commodity, LNG carriers are getting bigger. On order are a score of vessels of 210,000 cubic meter capacity or better, priced as high as $260 million.[27] They are single-purpose vessels built to standards established by the International Maritime Organization (IMO) and certified by one of the international classification societies, organizations such as the American Bureau of Shipping, Det Norske Veritas, Lloyd's Register, and Bureau Veritas. Representatives of the chosen classification society visit the shipyard regularly, perhaps even daily during particularly critical times in the construction period, and keep a careful eye out to see that the work is up to code. (Of the various certifying agencies, CH-IV International consultant Jeff Beale says, "They're all credible, because they're always trying to one-up each other.") The international standards are very complete—page upon page of requirements—and cover everything from layout to containment design and gas detection and alarm systems, from stress analysis to welding and pressure relief systems, from fire protection systems to emergency shutdown systems and instrumentation.

Not only are the classifiers watching, but the owners' agent is equally attentive. He is concerned not only about safety, which is the classifying societies' main interest, he's also interested in efficiency and those little details, things such as corrosion protection, that mean the vessel will be able to work for as many as forty years.

Coast Guard personnel also to some extent oversee the construction of vessels that will be coming to the United States, and there are U.S. Coast Guard inspection offices in Tokyo, Singapore, and Rotterdam. Particularly in the case of a new design, inspectors from the Coast Guard will check out the ship as she comes to life. And finally, before her first trip into a U.S. port, any new carrier is given one more inspection for her certificate of compliance to do business in the United States. Beyond the physical check of systems, such as cargo safety shutdowns and gas detection and emergency shutdowns in the engine room, the crew is asked to go through some test drills to demonstrate their capabilities in matters such as firefighting and lifesaving. Says CDR Chris Oelschlegel, a retired traveling inspector for the Coast Guard, "It's not to say that the international rules and regulations aren't strict—they are—but we do a comprehensive examination."

Sometimes, Coast Guard regulations that were stricter than the international ones have subsequently been adopted by the IMO; for instance, some years back, international standards for steel composition in cryogenic applications were brought in line with U.S. standards.

Even once she's collected all the pieces of paper with their acronymic identifiers—COF, COC, SOE—a vessel will still undergo a safety and security screening by the Coast Guard prior to every visit to the United States.

LNG ships are required to have double hulls, just as most petroleum carriers have today, and all are required to have by 2015. They have two entire hull forms, one several feet inside the other, with seawater ballast carried in many separate compartments in the space between. The design of the two hulls is dictated by IMO regulation as well as by the ballasting requirements of the vessel, which maintains a constant draft both under way and as she loads and unloads. Automated or manual systems fill or empty appropriate ballast tanks to keep her on her lines.

Inside the inner hull are separate cargo tanks—four or five, usually—which are heavily insulated but not refrigerated. The LNG is usually separated from the ocean by at least nine feet of hull, insulation, and tank wall, and in many places more. It stays cold for the same reason that boiling water doesn't get any hotter than 212°F—add heat, and it just boils faster. The small amount of heat that comes through the insulation to the tank boils off just a bit of LNG, but the rest remains at –260°F. The boil-off is usually retrieved and burned for the ship's propulsion, although more and more, in recognition that the product is of more value than are alternative propulsion fuels, vessels will be set up to reliquefy boil-off and return it to the cargo tanks.

The ships are designed to protect themselves from tank leakage, because LNG spilled onto ordinary steel causes brittle fracture. Beneath the manifold connections there are drip pans of stainless steel, which can withstand cryogenic shock. These pans may contain six inches of concrete to encourage the quick vaporization of any leaks. The first LNG ship that Captain Dick Johnston sailed on carried rocks in the drip pan, just to add surface area for faster vaporization. LNG tankers are apt to have a rusty patch down the side beneath the manifold area; that's because they pump seawater down the side of the ship during loading and unloading in case there's a failure that allows LNG to flow off the ship. The running water provides a continuously warm surface on which the LNG may land—however cold the pumped water from the ocean might be, compared to the –260°F LNG, it's warm, and vaporization will take place quickly.

There are two common types of cargo tanks: the self-supporting spherical ones that many people associate with LNG carriers, known as Kvaerner-Moss, Moss, or ball tanks, and so-called membrane tanks, which are fitted to and supported by the shape of the ship's inner hull. There are two membrane designs, Technigaz and Gaz Transport, but they function similarly.

The self-supporting spherical tanks are made of aluminum (on all but two ships, which have stainless tanks), and provide no structural strength to the hull. The tank is only near the hull's side or bottom plating in three tangential locations. The sphere is supported on a single cylinder, like a belt around its middle, so the tank can expand and contract without interference.

Yet another design, the prismatic system, is rather like a cross between the other two. It uses self-supporting aluminum tanks like the Moss system, but they're shaped to fit the vessel's hull, as on the membrane ships. Only a couple of prismatic ships are sailing today; they're on the Alaska to Japan run. The system is more expensive to build, but LNG guru Bob Lakey believes the design to be superior: "They're very rugged—they can take the North Pacific, where the seas would cause sloshing problems with membrane ships." Dick Johnston is among many who also believe this system to be the best, but the prismatic ships are more expensive to build, and there are no new ones on order today. The tank system is being used, though, in storage ships for liquid propane gas.

The spherical tanker Louisiana *docking at the export facility in Dandier, Northwest Australia. Note the rusty patch down the side amidships where water flows during cargo loading and discharging.* COURTESY OF CAPTAIN DICK JOHNSTON

As for the Moss design, a major argument against it is that there is so much void space between the cargo tanks and the inner hull, requiring a larger vessel relative to payload. On the other hand, the separate spherical tanks handle the sloshing of partial cargo loads better than do the membrane tanks. The classifying societies are working on the sloshing problem for the membrane designs.

In laboratory situations, Moss tanks have demonstrated that they will develop a minor crack long before they rupture. Should a crack ever occur in use, the gas and temperature sensors required by regulation are designed to detect the leak quickly, allowing time for the ship to steam to the terminal to empty the cargo. More cryogenic-safe drip pans are located beneath the tanks to protect the vessel. After 800 ship-years of operation, the company claims an extremely low rate of "unscheduled off-hire," or unplanned downtime, only 0.44 days per year per ship, none of that ever caused by the containment system.[28]

The spherical tanks also have been shown to have great structural integrity and survival capability in collision or grounding. In 1980, the *LNG*

Taurus grounded outside Tobata Harbor, Japan, in heavy weather. She had extensive bottom damage but no LNG was released. The cargo was off-loaded and the ship was "gas-freed" and taken to the shipyard for repairs; no tank damage was discovered.

Construction of Moss-type vessels is easier than that of membrane tank–equipped ships because they can be built in sections, permitting concurrent construction of hull and containment. The tanks themselves are manufactured separately, shipped in, and set in place by crane.

However, even though they take longer to build, ships with membrane tanks have become the most common, and about four out of five of the new vessels on order will be fitted with them. Membrane tankers are smaller and lighter per unit of cargo than the Moss-equipped ships, and require less power, say their captains. It's certainly felt that membrane ships are better suited for the larger capacity many companies are demanding today.

Some say that because they have less "sail area" than do the Moss ships, they are somewhat easier to maneuver, and they offer better forward visibility for their helmsmen. According to Dick Johnston, who sailed LNG tankers as long as there were American-flagged vessels at work, all LNG ships, of either type, have significant sail area. "They're about like a square-rigged ship would be, same square footage per side."

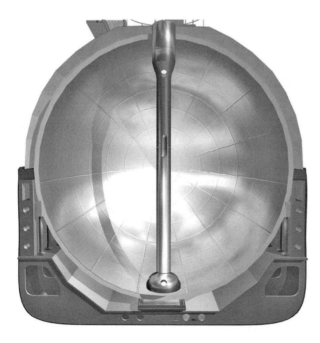

Cross-section of a Kvaerner-Moss tank. Keep the scale in mind; this tank is 125 feet in diameter. COURTESY OF SEAMANSHIP INTERNATIONAL LTD

LNG: A Level-Headed Look

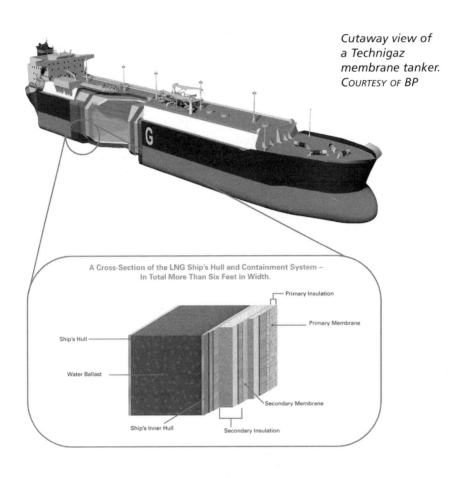

Cutaway view of a Technigaz membrane tanker. COURTESY OF BP

A Cross-Section of the LNG Ship's Hull and Containment System – In Total More Than Six Feet in Width.

Ship's Hull

Water Ballast

Ship's Inner Hull

Primary Insulation

Primary Membrane

Secondary Membrane

Secondary Insulation

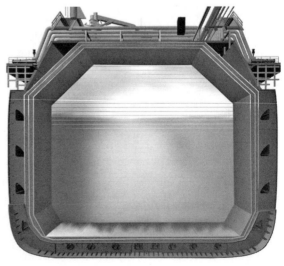

Gaz Transport membrane tank in cross-section. The ships are 145 feet wide or better. COURTESY OF SEAMANSHIP INTERNATIONAL LTD

One builder of LNG carriers, Japan's Mitsubishi Heavy Industries, builds both Moss and membrane vessels, and as of December 2005, has contracts for fourteen ships, nine of them Mosses. Their membrane ships use the Gaz Transport system of insulation, with primary and secondary membranes of 0.7-millimeter-thick 36 percent nickel alloy steel. There are two layers of insulation—historically in Gaz Transport systems, perlite granules in myriad separate wooden boxes, tied together with particular fittings—and the membranes are welded together. The whole construction is about twenty inches thick, and is very labor-intensive to construct, with more than 700,000 parts in the cargo containment system alone. "Each box is labeled exactly where it goes," says Ed Scott of Excelerate Energy, who has observed construction. "It's very impressive how they keep track of them all, so they go in the right place and they fit just so."

"There's an army of men in blue coveralls welding all day," describes civil engineer Chris Hosford, who also has spent time at one of the Asian shipyards. At night, another team comes in to x-ray the butt welds, as is required by the international standards. "Miles of x-rays."

The other common membrane-type containment system, Technigaz, has a waffled stainless steel membrane 1.2 mm thick and a secondary barrier of aluminum foil with strengthening glass cloth on both sides, along with foamed polyurethane or polystyrene insulation. The insulation for a membrane ship is expensive not only for its labor requirements, but it must be load bearing, as well, and it cannot be put in place until the hulls are basically complete, making construction a thirty-month project.

On Board an LNG Carrier

In 1986, after his sophomore year at Maine Maritime Academy, John Chapman shipped as a cadet with Energy Transportation Corporation's *LNG Virgo*, an American LNG carrier with a Far Eastern beat. (This vessel and her sisters were the last American-flagged ships carrying LNG. They were transferred to Marshall Islands registry in 2000, but still transport LNG around the Pacific.) The academy only requires students to spend sixty days working for a private company between their second and third years, but John wanted to get the most he could from the opportunity. For ninety-six days, he ran between Indonesia and Japan, back and forth, five round-trips altogether. As a cadet, he shadowed the merchant marine officers whose ranks he would join on passing his U.S. Coast Guard licensing tests. "I wasn't quite a grunt, but I had no authority. I made sixteen dollars a day." And two years later, in August 1988, he hired on with

ETC as third mate on the *LNG Leo*. His first tour was supposed to last 120 days, the typical run for the company, but it ended up a little longer: "Somebody had to get off early and needed a relief for forty-four days, and then they put me in a hotel for two weeks until the next ship came in, and I did eighty-six days on that." As the new kid on the block, the young third mate expected to pay his dues.

For a decade, four months on, more or less, and more or less four months off, he was in the "liner trade" between two LNG export plants in Indonesia and four import facilities in Japan. He sailed on six of the eight carriers that ETC owned, all built by General Dynamics in Quincy, Massachusetts, in the late 1970s. (No LNG ships have been built in the United States since 1981.) The ships were all of the distinctive Moss type, the domes over their spherical tanks protruding above the decks immediately identifying them as LNG ships. ETC's ships all had five spherical tanks 125 feet in diameter, and were among the earliest Moss-design carriers.

John describes the tanks as service-, maintenance-, and inspection-friendly. The interior of each tank could be accessed through a column that held all the piping and other equipment. When the vessels were in the shipyard for routine attention, crewmembers could climb down the ladder in order to change out pumps or make other repairs. They would get into the tank to inspect it, John says, and, for fun, "run in circles to try to get up the wall." The tank was so large in diameter and so smooth inside that they couldn't get very far.

The design permits easy atmosphere shifts within the tanks, which are always kept either too rich or too lean in their ratio of LNG to oxygen to permit burning. (Methane only burns within the relatively narrow range of 5 to 15 percent methane in air.) On the infrequent occasions when the vessels were in the yard and being worked on, the tanks would be open and filled with air. It would then take three days to prepare them for a new load of LNG, and much of the work was done during the southbound trip from the shipyard in Japan to the loading port in Indonesia. To preclude any chance of fire on loading, the first step in preparing to load cargo was to purge any atmospheric oxygen from the tank. An onboard inert-gas generator created a gas mix composed primarily of nitrogen, heavier than air, which they pumped into the bottom of the tank, pushing the lighter air out the top. The crew then changed around the piping of the cargo system so that when they got to the loading dock they could first put in a charge of natural gas (in its gaseous state) from the top. This incoming methane acted like a piston, driving the nitrogen

View from the bridge, the starboard side of the Louisiana. *The two insulated pipes along the top are for loading and discharging; one is for liquefied gas and the other is a vapor line. The smaller, uninsulated piping along the sides carries boil-off gas to the ship's engine room to be burned and is also used in exchanging air and gas when the vessel is in the shipyard.*
Courtesy of Captain Dick Johnston

mix out the piping at the bottom of the tank. Then the wrenches all came out again to change the piping once more. The heavier liquefied natural gas would then be loaded from the bottom and the gaseous methane would be vented out the top.

But the final filling of a cargo tank could only start after the tank had been precooled. They'd spray a small quantity of LNG around the inside of the tank to chill it before finally loading the cargo itself. And as the tanks chilled and contracted, all the fittings had to be tightened up. There was a lot to keep track of, but everyone involved understood the seriousness of their job.

On the return runs, they kept a small amount of LNG on board, perhaps 2 percent of the capacity, in order to keep the tanks cold until they were refilled the next time. If the tank is not already cold before being loaded, the incoming liquid can flash into a gaseous state and put too

The port side of the Louisiana, *photographed while traversing the Suez Canal. The pipes running up the side of each tank are part of the firefighting system. COURTESY OF CAPTAIN DICK JOHNSTON*

much pressure on the tank. If they encountered high tank pressures during loading, they had to slow the process.

Like those on all LNG carriers, the tanks on the ETC ships were neither refrigerated nor pressurized, but they were well insulated. The liquid stays close to –260°F on board, and keeps itself cold by very slowly boiling off. Under normal circumstances none is wasted. At the loading or receiving docks, the recaptured gas is simply sent ashore; at sea, on steam-powered ships like the ETC vessels, it is used to fire the boilers.

"Fuel oil consumption on the vessel could be two tons or ninety per day," says John, "depending on how much gas we had and how fast we had to get where we were going." When there was no hurry, they would allow the speed of the vessel to be controlled by how much gas was coming from the top of the tank. "There were times we were getting sixteen or seventeen knots just from that, if we had a hot cargo that had not been in the storage tanks in the load port for long and was weathering quickly." LNG turns to gas faster in rough weather, because sloshing exposes more liquid to the warmer gas above it. The unbaffled spherical tanks minimize the boil-off by leaving the smallest possible free surface at the top of the cargo.

Unless it was the first trip after time in the shipyard, loading took twenty-four to thirty-six hours, and then off they'd go to Japan. The voyage with product lasted six or seven days normally, but in different seasons, as Japanese inventories and demand varied, they might take fourteen days to make the turnaround, or even as long as a month. The first few years John sailed with ETC, they'd simply go to the anchorage at the loading port in Indonesia and drop the hook and hang out whenever they had time to waste. "The longest I was ever at anchor was fourteen days," he recalls. It sounds dreadful to be parked for so long, and he admits it was. "I must say, the company treated us well," he adds. "There were two or three launches a day, and we could go ashore." The company would plan baseball games between the ship's crew and the shoreside workers. "But," he admits, "there wasn't much there. It wasn't like going to a tropical resort—most Indonesians made less than $900 a year, and there was no nice hotel to stay in. You could get anything you wanted, but nobody had shoes, the streets were dirt, and gray water ran down the gutters." Once, John visited the Tanjung Puting Orangutan Rehabilitation Park, which he says was the highlight and about all there was to see. Normally, they weren't there long enough to do anything, anyway.

Later, when piracy became more commonplace, they dawdled under way rather than sit on anchor. "We'd make three knots during the daytime and ten at night," he says. "There were certain areas we didn't want to stay long." There were never actually any attempts at boarding during his time on the ships. "We were forty-three feet off the water at deck level. It's not like an oil tanker with six feet of freeboard. And we could move faster." The vessels could do an honest twenty knots on their own, and he reports making twenty-three knots with the current behind them off Mindanao, Philippines.

As third and eventually second mate, John was navigating through some of the most heavily fished areas in the world, off Japan and around Indonesia, Singapore, and Malaysia, where the fishermen work from canoes. The 936-foot ship is a large vessel to maneuver, and even though the bridge wing is 120 feet in the air, there's a long blind spot before her. In a storm or heavy rain, visibility is even worse, but they'd put a lookout on the bow, and accidents were very rare. John only heard of a single collision with a fishing vessel. The local people weren't the least bit intimidated by the huge ships, either. "We'd have people approach us, wanting us to stop so they could sell us oranges.

"You weren't going to see the world," he says. You knew what the

Aboard the Louisiana. *Above:* The manifold area, where incoming LNG is distributed to the piping along the tops of the tanks. *Right:* Looking aft toward the bridge. The dark windows of the bridge look small from this vantage point. BOTH PHOTOS COURTESY OF CAPTAIN DICK JOHNSTON

loading ports looked like and the receiving ports, and that was it. "But there were a lot of beautiful days out there, watching dolphins under the bow, and sunsets and coastal islands." He goes on, "And a lot of days I got seasick—everybody got a little bothered by going back out."

The hardest part of the job for any merchant mariner has nothing to do with what cargo they're moving. It always takes a few days of each tour to get back in the routine. "The first trip was pretty miserable," John recalls. "You were missing home, and you weren't quite in the rotation yet. Then you went on cruise control. It was easy until the last fifteen or twenty days, and then your mind was shifting to being ashore." The same pattern applied ashore: "You spend thirty days missing your lifestyle on ship, the independence. Then you're on easy street until the end, when you just don't want to be around."

John was married during the last three years he shipped out. "There

were times when my wife just hated it when I was given an extra two days. Sometimes we made the decision that I should go anyway. We'd already said our good-byes." Some mariners had trouble with their kids when they were home—they couldn't identify with the children, or they just played with them and took no part in discipline. Many of the men on ship were divorced. It was for these reasons that John chose to come ashore in the end; he took a leave of absence when his son was born, and he never went back. The money was very good—work in LNG paid higher than other shipping—but it wasn't worth losing his family.

He was never frightened by the cargo he was responsible for. The only dangerous times were when they were loading and discharging cargo, and most critical were start-up and shutdown. The chief mate was in charge, and the cargo engineer would be on hand during those times, but they could be relieved in between. It was a good source of overtime pay for John, who held a Person In Charge ticket, permitting him to stand watch on cargo control. This entailed monitoring and ballasting or deballasting to keep the vessel balanced as she lost or gained cargo. "It was straightforward," he says.

There were problems, yes, but procedures were in place to take care of everything. "There were safety systems and backups and routines, and we all knew what the dangers were. The potential of getting dusted in a car accident is much higher. The greatest risks out there were eating the local food or riding with an Indonesian cabdriver."

Unloading arms hooked up to the Louisiana. *COURTESY OF CAPTAIN DICK JOHNSTON*

The Louisiana *discharging cargo. The cloud behind the unloading arms is an ordinary fair-weather cloud in the sky, nothing to do with LNG.*
COURTESY OF CAPTAIN DICK JOHNSTON

Manpower and Flags

The safety of day-to-day LNG transport depends on experienced and knowledgeable seafarers, licensed and unlicensed alike. LNG ships are different enough from other vessels, and carry such valuable and potentially dangerous cargo, that they require the best of the best mariners aboard. An LNG ship is likely to have a crew of thirty on board, ten of them officers, and, of course, she has an entire second crew on holiday.

Today there is a worldwide shortage of merchant marine officers, a shortage that is only getting worse, according to President Len Tyler of Maine Maritime Academy. The LNG industry, in particular, faces a few years of serious shortage of qualified personnel. Traditionally, it has been a career job. A young officer might come aboard as a third mate and work for ten or twelve years, rotating on and off ship. "The company has a chance to screen for command, and that's why the industry has been so safe," explains Tyler. As of the end of 2005, about 180 vessels were in the fleet and at least 125 more under contract to be in service by 2009. During this same span of years a large number of the present LNG ships' officers will be retiring. The greatest number of LNG ships fly a Japanese flag—thirty-one, as of the end of 2005—and almost all of their senior officers will be retiring before 2010.[29] The same phenomenon is true elsewhere as well.

According to President Tyler, the industry may not be recognizing the severity of the shortage facing them. He serves as chairman of the International Association of Maritime Universities, which is working with the other training organizations, unions, and private companies with in-house training programs to develop solutions to the problem. SIGTTO, the Society of International Gas Tanker & Terminal Operators, has come up with a set of competency standards for all officer ranks on LNG ships for consideration by the International Maritime Organization. The IMO is also concerned that officers be able to receive and document the training they need. But there still will be a shortage of bodies. To become LNG certified, even a mariner who already holds an unlimited master's license may have to complete an additional 80 to 120 hours of course work, along with a certain number of cargo loadings and off-loadings. Fulfilling the latter requirement can be difficult, as the vessels don't have much capacity for extra personnel to be aboard just for training.

New licensed mariners are coming from Korea, China, Croatia, Poland, and other countries. "Hopefully," says Len, "at some point they'll come from the United States, too, even though there are no U.S.–flagged ships." There are people in the United States who feel that any vessel entering U.S. waters should have American personnel aboard, simply for security reasons.

Very little American shipping is left. Captain Larry Wade is master of the Maine Maritime Academy training ship *State of Maine* and in the business of turning out mariners. He bemoans the state of the American merchant marine. "When I got out of school, in 1964, there were 2,800 American-flagged ships," he says. "Now there are 400, and half of those belong to the government." Not one LNG carrier is in the lot.

During the sixty years since World War II, when we were one of the major maritime powers, we've lost more and more of our merchant marine, and many feel we cannot now be considered a maritime power at all. Two percent of the goods coming into our ports come by American vessels. To fly the American flag, a ship must be built in the United States, and it is no longer economical to build here without subsidization. "We taught the Japanese and Koreans how to build ships, and have since then forgotten how to do it ourselves on a commercial basis," says Ed Scott of Excelerate Energy, whose own history includes time in the shipbuilding business. "We still build the best ships in the world, but they have to be painted gray, and they're built with no budget constraints." As far as LNG carriers are concerned, Ed says, "We've been outspent on capital investment." In the early days of LNG tanker construction, American yards were on the cutting edge—the bleeding edge, Ed calls it—but when there was no money to be made in LNG, the motivation to keep up with the technology disappeared. Now, not one yard is capable of building an LNG ship competitively in the United States.

There's still interest on the part of some LNG importers in having Americans on board, according to MMA president Tyler. "LNG may be more acceptable here if some of the officers are American." But there aren't many qualified American mariners anymore, and many companies would not want to hire them anyway because of the cost. A container ship can be manned with a foreign crew for $5,000 a day, whereas a U.S. crew would cost the owner $17,000. Not all that differential ends up in the seafarers' pockets, either—income taxes and the various withholdings add up. Most other Americans working overseas get tax benefits that mariners don't receive. If they did, the pay would be more nearly comparable.

There are other issues, too. Sometimes our mariners are seen as "overpaid crybabies," in the words of one master mariner with a great deal of operational experience in foreign-flag vessels. More important, perhaps, we are seen as a litigious society, and the maritime system of compensation for injury encourages that view. Some feel that to revitalize our merchant marine, a good first step would be to replace the suite of personal injury, negligence, and general maritime unseaworthiness laws in the so-called Jones Act (1920) with a standard workers' compensation system. The Jones Act provisions support the lawyers very well but cost everyone else a lot of money.

Even American officers report this problem with American crews, which they say exists more among unlicensed seamen than those with licenses. Says Captain Dick Johnston, "Three times a year I'd have an insurance case. You ask a guy to spray some Windex, and he gets it in his eyes." Next thing, he's suing. "American officers and foreign crew would be ideal," Dick says. "Indonesians, Malaysians—you give those guys a room and a mattress and food, and they're as happy as can be, and they have money going home, too. They'll stay on board a year and be happy."

Realistically, the way most new officers will be found in the immediate future is by poaching from another company, Tyler says. Companies like BP that have various kinds of tankers may be able to swap out petroleum or LPG officers to take LNG ships. But somewhere down the line, there'll be a gap, and less qualified officers will be running tankers. He mentions a recent collision between a chemical tanker and a bulk carrier in the English Channel that may have been the result of human error. One might expect more such events with less experienced officers at work.

Shipping companies and maritime training schools around the world are working toward a longer-term solution. Companies are sponsoring cadets on their ships, to learn the ropes and get their sea time and required experience in, and they're contributing to programs to specifically train LNG officers in this country and overseas. ConocoPhillips donated $10,000 to SUNY's Maritime College to upgrade their simulator for LNG tanker training, and other companies are making similar contributions to other academies.

Simulators have been an important part of training for decades. They are a key part of the educational program in many of the maritime colleges and unions and in private companies as well. Marine Safety International is a private company with simulator courses available in Rhode Island, Virginia, New York and California. Their machines have the capability of modeling specific ships and can even re-create the cargo room of an LNG ship to train officers in the handling of cargo, as they did for Bob Lakey when he was responsible for designing and developing the marine facilities at the EcoEléctrica LNG import terminal. Lakey also arranged for the training of the Puerto Rican pilots. "You can put your own ship in, and the characteristics of the tugs you have available, so you can bring your ship in to the dock, punch button 'A' and turn the tugs on, direct them bow and stern, say 'push,' and it pushes." Even if there's nothing entirely like being on the ship itself, you can't just learn from the start on ships. "It would be like doing your driving training in a Rolls-Royce."

Bob says the educational facility run by Mitsui OSK Lines in Tokyo is extraordinary. "They could even train old Bob, I think." Mitsui OSK is the largest transporter of LNG in the world, operating sixteen ships currently, with another eighteen on order. "Management there is very dedicated to training," says Bob. "They don't want to be the ones to have the first bad problem."

Americans have found a few more opportunities in the engine room, says Len Tyler, because LNG ships are still steam-powered for the most part. It is only recently that the first diesel LNG ship was built. An advantage to steam is that the LNG boil-off can be used to power the vessel sometimes, not only making use of a fuel that otherwise might be lost but also providing cleaner operation in areas concerned about air quality.

Steam ships are also more reliable than diesels, with simpler engines that last longer and require less attention. The fuel costs are higher, however, which is more of an issue now than it used to be.

Most of the rest of the world hasn't trained a steam engineer in thirty years—all their ships are diesel nowadays—but the U.S. shipping fleet started the serious switch to diesel well after the rest of the world, and we're still turning out steam engineers. Maine Maritime Academy has kept its steam engineering program because its graduates are in demand for shoreside power plants. So there are trained American engineers who can go right to work on an LNG ship. On an LNG carrier, unlike most other tankers, the cargo has its own engineer; the ship's engineers don't need that cargo certification on their licenses.

It used to be that the so-called flags of convenience flown by many ships indicated that their safety standards and the conditions of their crews were questionable, but Captain Wade says that's no longer necessarily the case. Global standards for safety and training implemented in the nineties have brought seamen worldwide closer to a level of parity. It may still be easier to get licenses to work on ship in some countries than others, but these standards are being brought into line. Along with other security measures now being adopted worldwide, biometric identifications of mariners are required in more and more countries, an issue of particular interest in the United States, which is serious in its efforts to keep track of crews coming into our ports.

Building Safe LNG Carriers

In 2005, Tyler visited Daewoo, in South Korea, where there were seven LNG carriers under construction along with three or four Very Large Crude Carriers, a car carrier, naval ships, and a giant oil rig. He described the plant, which employs 40,000 workers, as being carefully landscaped and beautiful, and very efficient. "They're more productive than our yards, and it's not because the hourly wage is so much less than ours," he says. "They can build a ship in a third the number of man-hours than we can in the United States." The shipyards, built recently, are more advanced technologically than ours are, and they operate in a production-line manner. Our unions are part of the difference, he feels. "If a guy's a welder there, and there's no welding to be done, he's working with a screwdriver. No one's standing around." Even so, it takes a little more than two years to build a membrane LNG carrier, twice as long as for ordinary petroleum tankers.

Certainly a concern of everyone in the business is: What happens when one of these ships finds herself in trouble? While the industry's record to date is very nearly faultless, the simple matter of increasing numbers of vessels and numbers of voyages is necessarily going to stress the statistics. Perhaps, one of these days, from some cause, a cargo tank will leak. According to an engineer whose specialty is ship design, the first thing the crew would do in that case would be to flood the adjacent ballast tank with seawater. "It would ice up very quickly, stopping the flow," he explains. He believes that in the case of a small holing, depending on its location, there would be no fire. The standards require an inert gas (in this case, nitrogen) in the insulation environment. "As long as you have an oxygen content of less than 12 percent, no matter what the gas concentration, the insulation couldn't ignite." Even with a breach to the outside, the pressure of the inert system would preclude oxygen's entry, he says. There might be some brittle fracture if leaking LNG came in contact with unprotected steel, but he did not feel the ship would be in danger of being lost from such a leak.

But there are other situations that could lead to drastically different endings. In particular, the possibility of terrorist activity has to be considered. It's hard to conceive of an attack causing tank failure without a source of ignition. And then what would the consequences be? Many studies have considered the impact of a single spill, and models have been developed to calculate the size of a spill and the impact of fires from that release, but the question has been raised about whether the breach and subsequent fire would cause further tank failure.

The duration of a fire fueled by major spillage from a single cargo tank in an LNG carrier would be measured in minutes, not hours or days, says Dr. Jerry Havens, Distinguished Professor of Chemical Engineering at the University of Arkansas, who has probably pondered this and many other LNG questions as much as anyone. "Suppose it might be fifteen minutes. But the fire would probably be bigger than the ship itself. The bottom line is that it would not be over in such a short time that the ship would not be damaged."

The possibility of fires in multiple tanks must be a consideration when evaluating the hazards LNG can present. The insulation used in LNG carriers has come under much discussion of late. Some are concerned that it will burn, particularly if there were a gas leak from the cargo tank into the insulation, but that is very unlikely because of the inert gas system. While, in theory, the polyurethane or polystyrene insulation commonly used today can burn, a bigger concern is the fact that it melts or degrades even at temperatures too low to ignite it. Should that happen, the remaining cargo tanks would be left unprotected and, therefore, subject to failure, whether from overpressure or simple disintegration as the now uninsulated LNG warms and expands. The 2004 Sandia National Laboratories report sponsored by the Department of Energy, which is the LNG study that most people in the industry find most acceptable, gives credence to this view, concluding as it did that failures of three tanks would not be unlikely in a breached-tank situation.

Even if we knew accurately what the effects of a single fire from a single tank would be, which we don't, what would be the consequences of more than one tank spilling LNG? Would the fire increase in intensity or only in duration? (That would presumably depend on whether the tanks failed simultaneously or sequentially.) What about the cryogenic contact and brittle fracture as LNG dripped, drizzled, or poured from breached tanks? At what point would that alone threaten the integrity of the vessel herself?

It must be said that despite all the care that has gone into the design of LNG tankers, nothing is infallible. Not long after midnight on November 11, 2005, the *DBL-152*, a 442-foot double-hulled tank barge being pushed by her usual tug, ran onto a submerged oil platform gone astray during Hurricane Rita. Her outer and inner hulls ripped open as if attacked by a giant can opener, leaving a gash thirty-five feet long and six feet wide in her bottom. The full extent of the spill is not yet known, but the barge may have spilled into the Gulf as many as three million gallons of No. 6 fuel oil—thick, heavy, gooey stuff that sank to the bottom.

K-Sea Transportation, which owns the tug and barge, uses a satellite-based navigation system with routes stored in the computer in all its tugs, and the track taken by the tug's master traveling from Houston to Tampa was the very same that they had used for more than ten years. The immense oil rig is reported to have moved some 250 feet during the hurricane. Its location was apparently marked by small plastic buoys in the water, but no lights of any kind were working.[30]

Admittedly, the cargo of an LNG tanker is further contained inside the insulation layer within the inner hull, but it is not implausible that in such an extremely unlikely but not impossible accident, an LNG carrier might breach one or more tanks. The most probable scenario in that case would not have a happy ending for the ship herself, between what seems like inevitable cryogenic damage to hulls and structure and the likelihood of fire as the spilled LNG rose to the surface, and, most likely, ignited. Unless she were close to shore or other vessels, however, there's no reason to think the damaged tanker would put anyone at risk besides her crew and herself.

According to Alex Pasturhov, whose many years in the LNG business include time at the helm of the U.S. arm of the LNG tank-producing company Technigaz, "You cannot design for every disaster. It would be too expensive. You do that, you won't have any business."

The most impressive demonstration of the strength of tanks similar to those used to transport LNG came in 1974, when the double-hulled LPG tanker *Yuyo Maru No. 10* was T-boned in Tokyo Bay by the Liberian-flagged *Pacific Ares*, carrying 14,835 metric tons (MT) of steel products. The *Yuyo Maru* was equipped with several refrigerated tanks in which she carried propane and butane. The ballast tanks between her inner and outer hulls were filled with naphtha, itself a volatile, flammable liquid hydrocarbon. In effect, she was built much as an LNG ship is, although nowadays ships aren't permitted to carry any product in their ballast tanks. All told, the *Yuyo Maru* carried 20,202 MT of propane, 6,443 MT of butane, and 20,831 MT of light naphtha.

The forward section of the *Pacific Ares* was crushed in the impact, and at the point of collision, amidships, the *Yuyo Maru* was seriously holed, spilling a great deal of naphtha. According to a Japanese marine registry record, "As a result of the outflow of naphtha overboard, the sea surface on her starboard side literally turned into a sea of fire."[31] Both the forecastle and the bridge of the Liberian vessel burst into flames, as did the starboard side of the *Yuyo Maru*. All but one of the twenty-nine crewmembers aboard the *Pacific Ares* were killed; five from the *Yuyo Maru* died and seven others were injured.

An hour after the collision, tugboats arrived and began fighting the fires. The external fires aboard the *Pacific Ares* were extinguished within a couple of hours, but on the *Yuyo Maru*, the fire was more persistent. Several tanks of naphtha exploded, and smaller LPG fires burned from damaged deck piping and relief valves. For twenty-one days, the fires continued, and finally the ship was towed out to sea, small flames still burning. It then took more than a day's bombardment by four destroyers, four antisubmarine aircraft, and a submarine firing torpedoes to finally sink the ship.

The most noteworthy part of this entire tale is that in spite of the significant trauma suffered by the vessel, to the very end, the cargo tanks remained undamaged. Were it not for the naphtha, most likely there would have been no fires (and only seawater or air are permitted in the ballast tanks today). The ship herself stood up to all but the most vicious attacks by the Japanese Maritime Self-Defense Force. (It has been suggested that old Japanese submarine captains from World War II must have been amused, perhaps muttering, "These kids don't know what they're doing—we could do it in the middle of the night.") As the *Yuyo Maru* was very similar in construction to LNG carriers, she can serve as a demonstration of how much such vessels can take.[32]

To be fair, it must be said that though they are very similar, the tanks used for LPG are not identical to those which carry LNG. The insulation is different, for one thing, though many people don't feel that's significant to the comparison.

The *Berge Everett*

Owned and operated by the Norwegian company Bergeson, the big, all-green *Berge Everett* has been chartered for twenty years by Suez, the parent company of Distrigas, Everett, Massachusetts. She was delivered in the spring of 2003, and carries a bit more than 138,000 cubic meters of cargo. That's enough to keep more than 40,000 households in natural gas for a month. She makes regular deliveries to the Distrigas facility, but she also sails into other harbors when the opportunity arises. In fact, before she ever entered Boston Harbor at all, she'd already run between Qatar and Lake Charles, Louisiana, and had taken loads to Japan and South Korea.

The ship was built to accommodate as many different terminals as possible. She has a number of different landing platforms, making her compatible with seven loading facilities and a dozen import terminals around the world, and, as is required at some terminals, she is equipped with a system that monitors tension on her mooring lines. Her four membrane cargo tanks are insulated with perlite, a natural volcanic rock that has been expanded like puffed wheat, contained in plywood boxes. Loading and discharging are automated to a large degree, and cameras monitor all the operations. She can load a full cargo in about ten hours.

Like most LNG carriers to date, she runs on steam generated by fuel oil or LNG boil-off, or both, producing 37,000 horsepower that can send her along easily at twenty-one knots.

The automated engine room has more than six thousand monitored connections and alarms. Obviously, no one could oversee all this at one time, but an engineer can select one of fifty-five displays to observe at any of four screens. Quoted in a Bergeson publication, the *Everett*'s first master, Geir Harald Bratland, described her as being as close to a Cadillac as you can get. "She responds to all orders nicely and quickly," he said.

The photograph, which was taken from a residence in the old Navy Yard in Charlestown, shows the ship headed in through Boston Harbor. Visible are two of the several small guard boats that always accompany an LNG tanker making transit in the harbor. She has two tugs on this side, and, one assumes, another two on the other. The tug tied on forward is a traditional twin-screw diesel, and the newer tug aft is one of the tractors that are required for passage.

We can also see some of the external mechanisms of the ship herself. From the bow: the first mast is the standard mast any ship has, carrying the forward running light, a lookout station, and a fog whistle. The next two masts, plus two others, are cargo tank vents to release gas if need be.

There are two cranes, one on each side of the ship, to assist in picking up loading arms or hoses. The heavy piping along the deck leads from the tanks to the manifold, which can be seen on deck in the center of the vessel. On the roof of the pilothouse is an array of antennas for radar, GPS, satellite telephone, and probably television.

PHOTOGRAPH BY PATTY ARCUNI

LNG Characteristics and Hazards

IN its liquefied state, as in its familiar gaseous form, natural gas is color-less, noncorrosive, nontoxic, and odorless. It looks like water. It's the same stuff that many in urban areas use to heat their homes and cook dinner, but chilled to a temperature below –260°F, which is its boiling point. LNG weighs about 45 percent as much as water; regasified, it becomes lighter than air at –160°F. To show what it looks like, sometimes people put a couple of gallons in a foam beer cooler out in a parking lot. "It's like a pot of water you'd use to cook spaghetti, just starting to boil, with little tiny bubbles," describes engineer Neal Frangesh.

It may seem surprising that LNG is also nonflammable. LNG project developers like to talk about extinguishing cigarettes in liquid natural gas. (Of course, if one could somehow get a lighted match into a tank of gasoline, it would be extinguished there, too. It's an unusual liquid that contains enough oxygen to burn.) It is only after LNG has been turned back into its gaseous state that it becomes flammable—and then, only if it is at the right concentration in air, between 5 and 15 percent. The circus trick has gone awry from time to time, as it did in a school lab demonstration that Texas A&M adjunct professor Dr. Harry West recalls. They set the beaker in a hood and turned on the vent, which increased the vaporization rate, and as the cigarette approached, bingo! Fire.

One of the difficulties in speaking of the hazards of LNG is that its safety record has been so good. Very few LNG accidents have happened from which to build a relevant database. The physicists and engineers and chemists have studied what incidents have occurred, as well as a few somewhat similar cases involving other fuels, and have performed what experimentation they reasonably could. They've calculated and recalculated and analyzed each others' calculations, trying to predict just what would happen in this or that scenario involving LNG ships or facilities, but the bottom line is, no one has actually seen what might be possible because very little has ever happened. This is both good and bad—perhaps a sterling record in the past would suggest that such a record might continue, and, of course, all hope it will. But in the absence of a big event, it's hard to know for sure what the consequences might be.

The most recent figures I've heard account for more than forty-five years of LNG shipping around the globe, 100 million miles,[33] with a few accidents but never a major spill. Within the next very few years, more than half again as many LNG ships will be working as there are today. Sooner or later, it wouldn't be surprising if there were a more significant accident, just from the sheer numbers of voyages, even though there's no doubt that the industry is very cautious, indeed. (It's said that you're more likely to die in a plane accident approaching an airport in the middle of the desert with no hazards nearby than coming into Washington, D.C., or Boston, where you're dodging buildings and making sharp turns. At congested urban airports, the pilot knows he has to be on guard and he's extra cautious. Perhaps the same may be said of mariners carrying LNG compared to those whose vessels are loaded with gravel or timber.)

Every past LNG incident has contributed to the planning for future construction and operation of LNG facilities and vessels; each identified problem has inspired further honing of the applicable regulations and designs so as to minimize the likelihood of that particular difficulty arising in future. Too many lives and too much money are at stake for anyone to take safety lightly.

A key point is that much of this LNG history took place before September 11, 2001. Again and again, I've heard people in the LNG world say the rules changed entirely that day. The fact that there has been no terrorist attack yet on an LNG ship or facility cannot be interpreted to mean that there never will be one. Elsewhere in this book, we'll consider the likelihood of an attack; in this chapter, we'll look at the possible consequences of an accidental spill or a purposeful assault on an LNG carrier or facility.

LNG Spills and Vapor Clouds

The most-often raised questions about LNG regard the safety of its handling, transportation, and use. Of course, it is not entirely free of hazard. Nothing that burns is—not gasoline, nor coal, nor even cooking oil. "But some are worse than others, and liquefied natural gas is far from the worst," says Dr. Alan Schneider, a chemical engineer whose career with the Coast Guard began with LNG in 1974. He explains: "Some things detonate in an unconfined state—acetylene, propane, ethylene. Methane does not. Some things have lower auto-ignition temperatures—hexane catches fire when it reaches five hundred degrees Fahrenheit, and methane at one thousand degrees. Some things burn hotter—hydrogen burns hotter than methane, which is why the rocket to the moon was powered by hydrogen and not LNG." Acetylene burns hotter than methane, too. In addition, it is far less stable than methane due to its particular chemical structure. It also has a much wider flammable region—it burns at concentrations ranging all the way from 2.5 percent acetylene in air to 85 percent. Methane only burns within the relatively narrow range of 5 to 15 percent in air.

One of the dangers of LNG is its frigidity; it can freeze flesh, and it causes most metals, including steel, to become brittle. When LNG spills, it is so frigid that it condenses moisture from the surrounding atmosphere, and the resulting gas-and-vapor cloud hugs the earth's surface like a ground fog. The cloud would probably asphyxiate any unfortunate soul within it, or certainly cause lung damage and frostbite. LNG facility exclusion zones are designed to ensure that no one, save perhaps a plant worker, could ever find himself in that situation, however.

When liquefied natural gas runs onto land, it spreads until it forms a lake bounded by its surrounding terrain. For safety, LNG facilities are required to have an enclosed area of volume sufficient to contain what's stored on premises should there be a failure. Often, they are surrounded by a berm. If a tank were to fail, allowing LNG to spill onto the ground, there would be an initial flash of evaporation, the ground beneath the pooled LNG would freeze, and then vaporization would slow.

However, as LNG gradually warms, it expands. At its boiling point, the newly formed gas has already about 250 times the volume it had as a liquid, so the resulting vapor cloud will overflow the bermed area.[34] Where the vapor goes next depends on the air currents; it will drift downwind and downhill at about the speed of the wind. The more turbulent the wind, the more quickly the gas mixes with air and the more quickly it dissipates altogether, if it doesn't catch fire first.

On water, an LNG spill creates a convection cell, and evaporation goes faster. The frigid LNG lands on the surface water, which chills and sinks, allowing warmer water to rise to the surface in its place, further vaporizing the LNG. Because LNG is lighter than water, it simply spreads out over the water. The evaporation rate is influenced by wave action as well as wind speed.

On land or water, LNG vapor warms primarily by mixing with air, and the cloud remains denser than the air surrounding it until it mixes to the point that it is too dilute to burn. Absent ignition, LNG evaporates, leaving essentially no residue—a self-cleaning fuel spill.

When a spill first occurs, the gas-to-air ratio is too rich to burn. "As it mixes with air, only part of the vapor cloud is in the flammable range at any one time," explains Dr. Schneider. An LNG vapor cloud is generally flammable only near its edge; that's where it has reached the critical ratio of 85 to 95 percent air and 5 to 15 percent natural gas. In the interior of the cloud, there is not enough oxygen for ignition, and air can't penetrate it to raise that ratio. Outside the cloud, too little gas remains; the mixture is too lean to burn. Only the boundary is ignitable. "For that reason, an LNG vapor cloud cannot burn all at once but burns only as air mixes with the LNG vapor," Schneider explains. "It burns as oxygen diffuses into the over-rich LNG vapor." Combustion might occur immediately above the spill pool, or perhaps at some distance, as the cloud of vapor drifts.

Realistically, even if there were no immediate fire after a spill, an ignition source is likely to be close at hand. Machinery, a cigarette, even static electricity could serve. In fact, years ago, the suggestion was made that small flares be burned along LNG facilities' property boundaries in order to assure that vapor never escaped. Though the proponent of this idea was ridiculed by the industry at the time, his suggestion has, in fact, been used. Rob Wyatt was the permitting expert on the EcoEléctrica facility in Puerto Rico, which is on a point connected to the main part of the island by a narrow neck of land. At the final public hearing, asked yet again about flammable vapor clouds, he pulled out the site plan drawing. He explained about the natural gas flare at the edge of the property boundary. The flare's pilot light, just like that on a gas stove, creates a self-immolation flashpoint for any vapor cloud that drifts near the site property border. "The bottom line is, we are going to burn ourselves up before a vapor cloud ever even gets close to going offsite," he told his audience. ("Never was asked that question again," he recalls.)

This protective technique is mandated now around Texas salt domes used for natural gas storage. If there's a break in the piping below ground, the gas will percolate up through brine ponds, so they set electronic flares around those ponds. It is considered better to light leaking gas and let it burn in a known, controlled location than to let it travel.

The Sandia Report

So, how widespread would the thermal impacts of an LNG fire be? Again, as there has been no big spill to observe, the models are based on extrapolation and interpretation from empirical studies of smaller events, and people don't always agree on their interpretations. The studies use differing variables as well as differing rationales. In December 2004, the Sandia National Laboratories released the document now most frequently referred to by those wishing to consider the impacts of LNG spills on water. Entitled "Guidance on Risk Analysis and Safety Implications of a Large Liquefied Natural Gas (LNG) Spill Over Water," it is simply known as "Sandia" now, as are the laboratories themselves. The Sandia labs carry out a wide variety of studies for the Departments of Energy, Defense, and Homeland Security, and employ eight thousand people in all manner of disciplines—materials, explosives, fire science and modeling, cryogenic and structural mechanics, risk management, environmental issues, terrorism, safety—the list goes on and on.

Staff members with expertise in all those areas contributed to the Sandia report, as did LNG shipping consultants, members of the intelligence community, and others. Mike Hightower, who has been at Sandia since the late seventies, took the lead on the report. Whenever a particular issue came up, Mike had the resources at hand.

The Coast Guard and FERC reviewed the report, and it was subjected as well to technical review by an external peer panel. Mike stands behind the Sandia report both because of the depth of the resources available and because of the review process. "There's a lot of drive-by analysis being done," he says. "Guys saying 'I have a PhD, and I did a report in my spare time and I think this . . . .' This report is setting a foundation—if you want to discuss LNG, do it in a qualified and technically reviewed venue." Appendices make up fully half of the Sandia report, and give insights on much of the in-depth thinking behind the recommendations and considerations in the earlier sections.

The Department of Energy's Office of Fossil Energy commissioned the report with several goals in mind. The first was to provide an idea of the scale of impact in the case of an incident involving an LNG carrier. Reports from the 1970s had suggested that a fire or vapor release from an LNG tanker could result in harmful consequences scores of miles distant; more recent reports lessened the distances of impact, but still offered widely varying estimates. "Was it a thirty-mile problem?" Mike asks. "Was it ten miles? Or five?" Upon in-depth review of four recent studies, and much consideration of various scenarios, the Sandia scientists concluded that the scale of impact was more like a mile or two, depending on the size of the ship, the location of the event, and the conditions surrounding it.

The second goal was to provide a consistent baseline approach to analyze and manage threats, hazards, and consequences of an LNG spill over water. "It gave everyone a way to use the same terminology, look at the same things, make their evaluations in the same way in their risk management and risk mitigation," Mike says. "It's a question not of looking at the worst-case scenario, but determining what actually could happen, how can we prevent it, what can we live with, where should we site facilities so as not to be detrimental to the public health? We do risk management all the time in industry; here's how you do it with LNG."

The study concluded that the risks from accidental spills are small and manageable with current safety policies and practices. The record is certainly supportive of this conclusion. Even the risks from intentional events, says Sandia, "can be significantly reduced with appropriate security, planning, prevention and mitigation."

The most serious accidental tank breach they found credible would be from a waterline-level hole two square meters in size. A spill of half the cargo from one tank would ensue that could lead to a pool fire a little larger than 200 meters (656 feet) in diameter. Depending on wind and sea conditions and the rate at which the product flowed from the cargo tank, it would burn for twenty minutes before running out of fuel. Second-degree burns could be experienced 2,500 feet from the center of the fire, just under half a mile, and structural damage would be likely within 800 feet of the vessel.

Sandia also considered a range of intentional attacks on LNG ships in order to define credible threats. They consulted with intelligence sources about likely sizes of terrorist groups: possibly four or five men would be working together, perhaps in conjunction with another one or two similarly sized groups. They looked at the kinds of explosives terrorists would be likely to use and the kind of technical knowledge they would have—there's no reason to expect them not to be proficient technologically and have capable engineers working with them. The researchers continued with this sort of analysis and came up with a variety of scenarios they found credible, ranging from single-tank breaches to failures in multiple tanks, with hole sizes ranging from two to ten square meters.

The nominal case Sandia considered was an intentional spill and subsequent fire of 12,500 cubic meters of LNG, half of a ship's normal cargo tank, from a single hole of five square meters. The idea is that a deliberate breach would probably be made close to the vessel's waterline, and about half the contents of the tank are above the waterline. (Frustratingly, what's supposed to happen to the other half of the contents is not explained.) Such a fire would burn for about eight minutes, they calculated, and second-degree burns would be experienced within a radius of about 4,300 feet (eight-tenths of a mile).

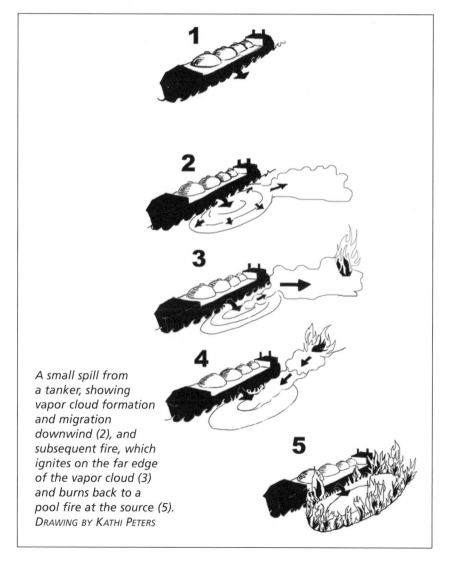

A small spill from a tanker, showing vapor cloud formation and migration downwind (2), and subsequent fire, which ignites on the far edge of the vapor cloud (3) and burns back to a pool fire at the source (5).
DRAWING BY KATHI PETERS

One of the interesting aspects of the Sandia report is the conclusion that a three-tank fire is the most plausible multi-tank event. This is not, as the report implies, because only three tanks could be damaged as a result of an attack, but rather, Mike explains, that at any given moment, there would be little likelihood of more than three burning simultaneously. The fire from a single tank with LNG flowing from a hole of five square meters in size would be expected to last only six to eight minutes. During that time, the insulation on adjacent tanks would likely be compromised, and there might also be cryogenic effects on structural members of the ship, either of which could cause failure in one or two more tanks. They could then contribute to the fire—but the first tank would be running out of fuel by that time. The next failed tanks would affect their neighbors in the same manner and time frame, leaving but three likely to be burning at any given time, Mike and the other authors of the report concluded. In that case, the most extreme impacts they foresaw were a pool fire of almost 1,900 feet in diameter, property destruction out to 2,000 feet from its center, and a distance to second-degree-burn level of nearly 7,000 feet (1.3 miles).[35]

The distance that a natural gas vapor cloud can migrate and still remain flammable has been the subject of much discussion over the years. It was suggested in the 1970s that flammable vapor might be able to travel as much as fifty miles or maybe even a hundred, but those distances have long since been dismissed by engineers and scientists. Although it is very difficult to imagine how tanks behind two hulls and a thick bank of insulation could be holed without a source of ignition, in order to get the largest vapor cloud drift that they could, the Sandia team assumed that somehow the breach took place in three tanks simultaneously from holes totaling five square meters, and with no fire. For the same reason, they chose a low wind speed of two meters per second (a little under four and a half miles per hour); faster would simply disperse the gas more quickly. This was the worst scenario they found to be even remotely credible.

Mathematical models have been developed to simulate the atmospheric dispersion of such clouds. When Sandia plugged in the variables, they determined that their theoretical spill would cause a vapor cloud that would wander off downwind, reaching a gas concentration lower than the flammable limit almost 12,000 feet away, two and a quarter miles.[36] At four and a half miles an hour, that journey would take about half an hour.

Naturally, an actual incident would be somewhat different from the models Sandia considered, depending on the vessel's design, the weather and sea conditions, the local surroundings, and so forth. The work did establish, however, a reasonable scale within which to consider consequences, as well as a framework within which to consider specifics.

The last objective of the study was to develop guidance on risk management of LNG operations on the water. Identification of the risks and their possible consequences is essential; the report gives examples of possible hazards and considers strategies to reduce or manage each. Prevention of trouble is best; mitigation after the fact is important, too. As examples of potential threats to LNG carriers, Sandia considers ramming, triggered explosion, hijacking, and terrorist attack with explosives. For each of these scenarios, they chart local and cascading consequences and risk reduction measures, both preventive and mitigating. For instance, to minimize likelihood and consequences of a hijacking by insiders, they suggest early interdiction and searches, control and surveillance of the ship, and employee background checks as preventive measures. To mitigate the consequences of a hijacking, they list an emergency response force and evacuation plans, both premeditated plans of action.

The Sandia report has become the most quoted and most referred-to study of the many published LNG safety analyses. It is seen as being objective, and for the most part it has been accepted as fulfilling its intended purpose: to provide guidance in planning for the safety and security of LNG carriers as they go about their jobs.

LNG Pool Fires and Other Phenomena

Fires from spills of LNG on land and water would normally burn out quickly—in minutes, most likely, unless there's only a small leak from a large tank—but they do burn with intense heat, hotter than gasoline or oil fires of equal fuel size. A burning pool of LNG spreads until it's burning off at the same rate that the fuel spills. An LNG fire of any size could certainly kill or seriously burn anyone near it, as well as set flammable objects on fire.

In a city, the very likelihood of a natural gas cloud's running into something that can set the vapor ablaze might be that city's salvation; once the cloud has caught fire, it simply burns back to the spill, its fuel source. There's no pressure wave, no boom, and there's no high-speed movement of flames. The flames will eat their way back into the cloud at the rate of about six feet per second, or four miles per hour, just a little faster than an ordinary walk, stopping when it runs out of fuel. This has been demonstrated in test after test, starting in China Lake.

Back in 1979, the Coast Guard joined with the Gas Research Institute to perform some experimentation with LNG fires at the China Lake Naval Weapons Testing Site in the deserts of California. They discharged a few cubic meters of LNG for each test, creating a 10-meter (33-foot) diameter pool. At those small sizes, the fire burns clean and clear, yellow and white like a Bunsen burner, just flames and little smoke, and very hot indeed. (By contrast, even a small oil fire burns with a lot of billowing black smoke, which is unburned carbon sailing off without emitting much energy as heat. It's not as hot.) But with a larger LNG spill, black smoke appears. When a 35-meter (115-foot) diameter spill was tested in France, the whole fire was orange and black. It looked and behaved more like any other hydrocarbon fire. The large fires are sooty, it is surmised, because there is insufficient oxygen at the center of the fire. That soot also blocks heat radiation, making the size of the hazardous area smaller than a simple extrapolation from the smaller tests would have suggested. It has even been posited, though not yet proven, that the emissivity of LNG fires will not increase much from pools of increasing diameters beyond a certain size.[37]

What about explosions? One of the objectives at China Lake was to investigate the likelihood of unconfined vapor cloud detonations. Try as they might, they found that they couldn't create an explosion with gasifying LNG in their open-air test conditions unless it contained a significant proportion of propane, which *is* known to explode in the open. At 86.4 percent methane and 13.6 percent propane, the gas mixture wouldn't explode; at 81.6 percent methane and 18.4 percent propane, it would. Somewhere in between is the cutoff.[38]

Even though other experiments since that time have consistently failed to make LNG blow up unless it's in a confined area or contains high proportions of heavier hydrocarbons, some credentialed engineers state simply that any hydrocarbon can be made to explode, given the right conditions. After working in oil and gas for some thirty years, first for ARCO and then as an independent consultant, Richard Kuprewicz believes that. "I've seen it," he says, explaining that in the course of his career he has personally observed situations that demonstrated that unconfined LNG vapor *can* explode. He believes the reasoning offered by FERC and the consultants working hard to get LNG accepted to be biased.

"There are many operators who can't speak up that have had near misses but can't be candid about it for fear they're going to lose their jobs or their pensions." He himself is bound by confidentiality, he says. "Have I experienced an LNG release? Yes." That's as far as he'll go.

Kuprewicz dismisses other consultants for not having actually been operators of LNG facilities themselves, although certainly there are today many former operators working as consultants. One is Arthur Ransome of CH-IV, who has been in various aspects of oil and gas as long as Kuprewicz and who ran the EcoEléctrica LNG import terminal and electrical generation plant in Puerto Rico for a number of years. Ransome clearly believes in the safety of liquefied natural gas. Asked outright whether he is more comfortable with it now than before he worked directly with it, he says unequivocally, "Yes."

There is also Dr. Alan Schneider, who was one of the lead engineers on the China Lake tests. He had and still has no axe to grind about LNG because he has no stake in its being accepted or not accepted. He simply says that if theory and experimentation or experience conflict, the burden of proof is on the theorist. Under experimental conditions in open air, vaporized LNG has not been made to explode, and no one has demonstrated that it has ever done so in the real world, either. It seems reasonable, he says, to conclude that it cannot explode in unconfined spaces. Within the vapor cloud, only a small amount is at the right range to burn at any given time. On the other hand, if a proportion of heavier hydrocarbons such as propane are in the mix, as the lighter methane evaporates off, the heavier gas may constitute a greater share of what's left, leaving the possibility of explosion.[39]

No one disagrees that a burning vapor cloud is serious enough to those within or near it even without detonation. The job of everyone concerned is to prevent an LNG spill in the first place.

So, even if LNG doesn't explode in unconfined areas, are there other particularly dangerous possibilities?

A fireball can develop in an LNG spill situation. Fireballs are certainly dramatic, shooting off as much thermal energy as a pool fire, but they are only possible when a quantity of fuel and air mix together before ignition, so they would not be expected to be common with LNG. They are of extremely short duration—seconds to tens of seconds—and, according to Sandia, are of relatively low threat.[40]

Another possible hazard is called a Rapid Phase Transition (RPT), an explosion-like event, but without a fire. When a quantity of LNG falls on or into water, heat transfers from the water to the LNG so quickly that the LNG instantaneously converts to a gaseous state, and *BANG!* The fast vaporization appears to be an explosion, but there is no fire and no heat, and while the blast can be a hazard to anyone or anything in the immediate area, the energy released in an RPT is considerably less than that from combustion. The shock wave from such events would only be felt near the spill source, and shouldn't cause extensive structural damage. [41] This is, however, one of the areas that some feel warrant further study, as perhaps RPTs could cause secondary damage that would lead to other failures of facilities or tanks. If a tanker were to be breached from underwater, perhaps an RPT would occur. That said, the cryogenic damage to the vessel might well be greater than any it might suffer from an RPT. Based on his and others' observations of experimental RPTs, Dr. Schneider says that they do not seem to get bigger as the spill size increases.

Project managers promoting the safety of LNG used to like to pour some into a fish tank to show that it doesn't harm the fish, until one such demonstration went amiss. In the words of Dr. West: "A little bitty RPT blew the aquarium apart." One can conclude that the fish were indeed hurt by that.

Another issue that is still under discussion is the Boiling Liquid Expanding Vapor Explosion (BLEVE). The acronym is turning into a word, pronounced "blevy," and it is used as a verb as well as a noun. A BLEVE can occur when a tank containing a pressurized, volatile, flammable liquid is exposed to heat. One often hears that since LNG isn't stored under pressure it can't BLEVE, but this is questioned. As a liquid-filled tank heats up, the contents boil into a gaseous form. The gas requires much more volume than did the liquid. (In the case of LNG, it's usually reckoned to be about 250 times more in volume.) If there's no release possible, or the relief valving isn't adequate to handle the expansion, the tank explodes. It's a very dangerous situation, causing not only a potentially huge fire, depending on what's in the tank, but shrapnel flying great distances as well. But because LNG is carried and stored in unpressurized containers, many engineers believe that an LNG tank would simply fail without blowing itself up, and that an LNG BLEVE, if it were to happen, would not be as forceful as a BLEVE of a pressure tank.

No one can truly say what the actual odds of problems are with LNG. Shy of terrorist attack, which we'll come back to in a later chapter, the industry is very confident that its good record will continue. But any trouble would be serious.

"What I get from those models is more limited," says Dr. Schneider. "Any effect is sufficiently bad that the Coast Guard had better prevent disaster or fire. And our people think we can."

LNG-Related Accidents
(And Those Blamed on LNG)

YOU can't talk about LNG for very long before someone mentions Cleveland. The 1944 fire at the peak-shaving plant in Cleveland killed 128 people and injured 255 more; three-quarters of a square mile was involved, and everything on about thirty acres was destroyed, including seventy-nine homes and two factories. That disaster initiated a twenty-year slowdown in LNG development in the United States, but it is not happenstance that there has been no incident even remotely near the scale of Cleveland ever since. People learned from Cleveland, and with each less disastrous event since that time, lessons have been searched out and safety requirements codified.

Natural gas is dangerous not only because it can burn, which indeed it has to do or no one would have any interest in it, but also because, in its liquefied state, it's so frigid that it can crack many tough-seeming materials. It was temperature-caused metal failure that precipitated the Cleveland disaster.

Commercial liquefied natural gas plants have existed in the United States since 1917, when the first was put into service in West Virginia. If there were deaths, injuries, or problems of any sort at any of them during the first twenty-seven years, they were not deemed significant enough to have been noted in any of the many compendia of information about LNG's history. The first event regarding LNG that anyone appears to have found noteworthy was the disaster in Cleveland in 1944.

For three uneventful years beginning in 1941, the East Ohio Gas Company of Cleveland operated a peak-shaving facility. In 1944, a new storage tank was installed. As stainless steel alloys were hard to come by during the war, the tank was built using an alloy with a lower nickel content than usual (one that has since been prohibited for this purpose because of its susceptibility to fracture at LNG storage temperature). Exposed to the product's extreme cold, the inner shell of the new tank became brittle. The problem was probably exacerbated by the plant's proximity to both a major railroad line and a plant that stamped out bomb casings; vibrations from those operations may well have accelerated the cracking. On October 20, 1944, when all the tanks in the facility were filled in preparation for the coming winter, the new tank failed and 1,100,000 gallons of LNG spilled into the nearby streets. Many possible ignition sources were nearby—machine shops, electrical devices, and so on—and the cloud of vaporizing gas caught fire. The liquid also poured into the sewers, where it ran for some distance before igniting. In these enclosed spaces, when the liquid vaporized, the gas did explode, cracking streets and weakening building foundations.

Although the initial fire was nearly out within twenty minutes, it weakened the supports of the closest of the original tanks, which toppled over and spilled a half-million gallons or so of LNG, which immediately began to vaporize and then ignited.

The incident spawned much thinking about LNG plant design. The leakage was caused in the first place by the inadequate steel used to construct the new tank. The materials now permitted for cryogenic use have been thoroughly tested and must meet strict specifications.

Next, the only explosions during the Cleveland incident occurred in confined quarters, within the sewers. The LNG flowed into the storm drains because the containment dikes around the LNG facility were inadequate. Designers today must take great pains to ensure that spill containment areas can accommodate the entire contents of the LNG storage tanks. In addition, the plant design must eliminate any conceivable ignition source within the containment area.

The third factor that allowed the Cleveland spill to turn into a disaster was the failure of the old tank's support structure, which was not insulated against the heat of the initial fire. Now tanks are spaced at greater distances and protected from fire. Lastly, the facility was located right in the middle of a mixed-use area, with residential housing adjoining it. Setbacks are now required.

The Bureau of Mines, which investigated the 1944 incident, concluded that with "proper precautions," liquefying and subsequently storing liquefied natural gas was "valid."

Designers of LNG facilities worldwide, to say nothing of their regulators, have taken the lessons from Cleveland to heart, and remarkably few significant incidents have happened since. In the more than sixty years since 1944, in more than two hundred LNG facilities worldwide, there have been fewer than a dozen serious accidents directly related to LNG, and most were due to human error. None was caused by tank failure.

The worst event since Cleveland took place in 2004 in the liquefaction plant at Skikda, Algeria. As of the winter of 2005–06, a final report about the causes of the catastrophe had yet to be released. The root cause appears to have been a badly maintained steam boiler, although a gas fire was responsible for some of the deaths. Unlike an import facility, the liquefaction plant uses boilers to create steam for turbines to generate power to run the refrigerant compressors that liquefy natural gas. In the early evening of January 19, 2004, one of those boilers blew up. According to a report at the time, the boiler had been known to be defective for over a year and only superficially repaired.[42]

The explosion took place between two sections of one of the facility's six "trains," as the LNG-producing units are called. At different sections within a train, the heavier propane and ethane gases are separated from the methane, then the methane is liquefied. The pressure within a steam boiler was rising, and despite the operator's attempt to reduce the fuel flowing in, its safety valve opened. At the same time, a hydrocarbon leak was causing a gas vapor cloud to form nearby. (No one has yet said what piece of equipment had failed or why, nor which type of gas was vaporizing.) An air-inlet fan pulled the leaking gas into the firebox of one of the boilers, where it reached the right mixture with air and exploded, thereby causing a second, much larger explosion of the rest of the semi-confined vapor cloud within the building.[43]

Three months later, Sonatrach, the Algerian government–owned oil company that owns the liquefaction plant, published preliminary conclusions, saying it was "difficult to find at this stage which fluid leaked and from which equipment or pipe due to the damage level in the train explosion area."[44] Engineers are divided on their thinking; some believe that it was a leak of propane or another heavy (and known to be explosive) hydrocarbon gas. Others, however, are convinced that the accident did involve LNG. The fact that there is no final report yet, says Richard Kuprewicz, is in itself evidence that LNG was involved: "Someone's covering something up."

Twenty-seven people lost their lives at Skikda (most due to the blast, rather than by fire), and another fifty-six were wounded. All were employees of the plant. Three of the trains were destroyed in the fire, at a cost of about $800 million.[45]

The third accident most often mentioned in conjunction with the dangers of LNG took place in 1973 in an out-of-service LNG tank in a peak-shaving facility on Staten Island, New York. The tank, which had been leaking for well over a year before it was emptied, was warmed and purged of combustible gases. There was certainly no LNG remaining in the tank, but the Mylar tank liner was ripped in places. Ten months later, repairs and cleanup were under way and while workmen were in the tank, there was a fire; pressure built up in the tank and blew off the roof, which then fell on the forty workers below. The medical examiner stated that the men died of asphyxiation; the fire had consumed all the oxygen within the tank.[46] (Some reports simply stated that the roof fell on the workers, killing a number of workers. The number of reported fatalities also varies.)

The New York Fire Department investigated the accident at the time and called it a construction accident, not an LNG accident.[47] The Federal Power Commission also reviewed the event, having found that there were discrepancies in the construction plans and materials as well as in operations, particularly as to safety. They concluded that natural gas had seeped through the liner and the insulation surrounding it and into the earthen berm around the tank. Perhaps some was entrapped in the insulation or surrounds. Another explanation is that it was fumes from a cleaning solvent that ignited.[48] In any case, something set the Mylar afire. The workmen were not dressed in cotton clothing and slippers to reduce static electricity as they had been instructed, and, despite orders not to bring flammable or metal objects into the work area, a number of potentially flammable items, including cigarettes, lighters, a gun, and highly flammable aftershave cologne were found in the debris in the tank. The Bureau of Mines did some testing at the time and concluded that a pressure rise from fire in the insulation alone would not have been sufficient to lift the tank roof, and that there also must have been some gas fed into the fire.[49]

In 1988, the New York Energy Planning Board concluded that, despite considerable investigation of the accident, information regarding the cause was inconclusive. "The most plausible explanation for the accident is that an unknown source ignited the cleaning vapors, which in turn, ignited the Mylar liner. The Mylar in its upright position burned much faster than in the tested horizontal position and caused the polyurethane foam to ignite. The resultant fire caused the temperature in the tank to rise, generating enough pressure to dislodge the six-inch-thick concrete roof, which then fell on the workers in the tank."

Perhaps the most significant finding made by any of the investigators is expressed in the final paragraph of the Energy Planning Board's discussion of the incident: "The government regulations and industry operating practices now in place would prevent a replication of this accident. The fire involved combustible construction materials and a tank design that are now prohibited. Although the exact causes may never be known, it is certain that LNG was not involved in the accident and the surrounding areas outside the facility were not exposed to risk."

It seems that the tragic Staten Island accident was a result of a series of avoidable errors (some of them even conscious). It is inexcusable that regulations regarding materials, construction methods, and operation management were ignored from the start to the finish of the tank's life, culminating in the decision to allow the leak to continue as long as it did. That the repairmen were neither adequately trained nor supervised is also hard to defend. This attitude of laxity can lead to disastrous results in any field. But no matter what, LNG cannot be blamed.

Perhaps the only fatality since Cleveland that was directly attributable to LNG took place in Algeria in 1977, when a worker climbed out to the end of a springy catwalk to adjust a manually operated aluminum valve on the ship loading line atop an in-ground storage tank. The valve housing ruptured, releasing a torrent of LNG over the worker, who died instantly. No one on the ship or in the control shack on land realized there had been an accident, and the spill continued until 1,500 to 2,000 cubic meters of LNG had been released—between 375,000 and 500,000 gallons. Then the man in charge of the ship saw a strange white cloud coming at him, threw the lines, and backed the ship away. The vapor cloud did not ignite.[50]

Such valves are now constructed of stainless steel and, at least in the United States, are not operated manually.

At the Cove Point Import Terminal in Maryland in 1979, there was one death, a serious injury to another person, and about $3 million worth of damage when LNG leaked through an inadequately tightened seal. It vaporized in its illicit travels along an underground electrical conduit and entered an electrical substation at the plant. Ignited by normal arcing contacts in a circuit breaker, the mixture of natural gas and air exploded. The National Transportation Safety Board concluded that it was an isolated incident and not caused by incorrect design or construction. Still, building codes were modified to prevent a repeat of that sequence of events.

Other deaths have occurred at LNG facilities, but in no way were they related to LNG. One took place in 1968, in Portland, Oregon: a construction accident killed four workers inside an unfinished LNG tank. In 1983, in Bontang, Indonesia, a simple error during maintenance caused failure of a heat exchanger, which blew to bits and killed three. There was no LNG present in either instance.

There have been other incidents at land-based LNG facilities, but a surprisingly small number, and none that led to anyone's death. In 1965, on Canvey Island, UK: a small spill ignited and a worker was seriously burned. In 1971, in La Spezia, Italy: gas "rollover" within a storage tank led to the release of a lot of vapor, but there was no ignition or injury. In 1972, in Montreal, Canada: an explosion in a peak-shaving facility involved neither LNG nor injuries. In 1978, on Das Island, United Arab Emirates: a spill from a failed pipe connection led to a vapor cloud, which dissipated, with no fire and no injuries. In 1987, in Mercury, Nevada: an accidental ignition at DOE testing grounds caused a thirty-second fire, with no injuries. A very few other accidental ignitions have taken place in large-scale tests, too, but none that caused injury. In 1985, in Pinson, Alabama: welds at a peak-shaving facility failed; LNG ignited, and six employees were injured. In 1988, in Everett, Massachusetts: 30,000 gallons of LNG spilled through a blown flange gasket. The spill was contained as designed, and there were no injuries and no damage, but operating procedures were changed thereafter. In 1989, in Thurley, UK: human error led to a jet of LNG being released and igniting, causing burns to two people. In 1993, in Bontang, Indonesia: an LNG leak underwent Rapid Phase Transition, causing damage but no fire or injury. In 2000, in Savannah, Georgia: a ship crashed into the LNG unloading pier; the plant was empty at the time, and there was "no indication that had LNG been present in the piping that there would have been a release."[51]

And that's the lot of them.

Of course there have been the usual maritime incidents also, the sorts of events that happen to any ship. In the course of scores of thousands of trips, there have been five groundings and a half dozen collisions and al-lisions, none of which led to any spillage of LNG. (An allision is a collision of a ship with a fixed object, like a bridge or a pier.) A couple of lightning strikes in the 1960s ignited vapor being vented during procedures that were routine at the time and are no longer; the fires were quickly extinguished by purging the vent with nitrogen. There have been a few minor spills during loading that caused chill-fractures in steel plating on the vessel but no other damage. Decks are now fitted with drip pans wherever leakage might occur. A couple of leaks while under way led to small amounts of LNG entering the inter-barrier spaces around tanks. None of these led to any significant problem.[52]

It's good to take a look from time to time at events that go just as they're planned, like one in Georgia in March 2006. When the *Golar Freeze* was discharging her load of LNG at Elba Island, she was suddenly yanked fifteen feet away from the pier by the surge from a passing chemical tanker. The emergency disconnect coupling on the unloading arms worked exactly the way it was supposed to, and there was no spill. The two tugboats on hand pushed the vessel back into place. Unloading did not start back up again for nearly a day and a half while the Coast Guard and an engineer from FERC checked everything out, but there was no harm done.

Trucks too have had their share of incidents, but there has been a re-markable safety record on the road just as on the sea. In fact, although LNG has been traveling overland since the 1970s, and the trucks have had the usual accidents—rollovers, collisions, wheel fires—it wasn't until June 2002, in Spain, that there was ever any kind of a fire involving the cargo. A road tanker rolled, burst into flames, and the pressure built up in its single-skinned tank until it blew apart. It's not clear if the initial fire was diesel- or LNG-fueled. It had been burning for nearly twenty minutes when the cargo tank burst apart; a white cloud appeared and ignited immediately into a fireball. The driver died, and two people standing a couple hundred yards away received burns. Parts of the truck ended up scattered more than 250 meters from the accident site.

There's been discussion as to whether the tank BLEVEd or simply failed, but in engineer Chris Hosford's mind, there's no question. There was no BLEVE, he says, "it was a straight overpressure failure." After thirty years in the field, including experience in cryogenic tank design, he's certain that the way the tank broke up and scattered precludes it having been a BLEVE: "If it had been a BLEVE, it would have gone like a rocket ship." He describes propane BLEVEs that threw their containers a mile and a half through the air. One 60,000-gallon propane tank shot two and a half miles, "and stuck in the ground like a rocket."

Whether this particular truck's cargo tank did BLEVE or not may be a matter of semantics; one can find different definitions for the term. More to the point, according to Hosford, trucks like the Spanish one have been outlawed in the United States for flammable-gas service. Its tank was a simple, single-shell vessel with flammable insulation and only a thin covering to protect it. "That insulation goes away in a fire," he says. U.S. trucks have an inner tank and an outer jacket of strong structural carbon steel, with six or eight inches of vacuum serving as insulation in between.

Perhaps even more important is the fact that the relief valves on the Spanish vehicle were much smaller than are required in the United States. Trucks in this country must have two sets of relief valves, big ones for emergencies, set to go off at seventy-five pounds of pressure, and small ones, for road release at fifteen pounds. The LNG industry likes to say that LNG won't BLEVE, but Chris makes it clear he's not saying that an LNG tank *couldn't* BLEVE. The key is providing adequate relief valves.

LNG did burn in a more recent truck accident in the town of Fernley, Nevada. The driver noticed a leaking valve and made it worse while trying to fix it. He called the local fire people, who, in turn, called the hazardous materials team from nearby Sparks. Captain Thomas Garrison made the incident report for the Sparks Fire Department.

"It wasn't anything very alarming," says Tom. "You could see a little liquid leaking, but when it hit the air, it would vaporize. Some hit the ground, but you couldn't see it, because it got a cloud above it. That was drifting with the wind—once it vaporizes, eventually it gets lighter than air."

There isn't much sign of civilization near the accident site—they evacuated a truck stop and closed off the interstate highway running nearby, and that left the area empty. Tom doesn't know what ignited the vapor. "It could happen a number of different ways. We haven't found the exact cause. Nobody actually saw it start. After the initial ignition, the fire burned off everything away from the truck, and it was contained at the rear of the truck. It was a *whoosh*, like a pilot light turning on your hot water heater, and the flames were twenty feet in the air at first, with a diameter of ten feet."

It was his team's first experience with a cryogenic tank fire, and they were concerned about a potential BLEVE. The firefighting team had been told that LNG tanker trucks won't BLEVE, but, understandably, Incident Command didn't want to take a chance. "A BLEVE is pretty impressive. That's why we were a mile away," says Tom. They wanted to be sure no one was hurt.

The fire never did get any more intense. "We watched it for about three and a half hours, and eventually saw the intensity diminish. If I had to play it again, I'd play it the same way—you're at the mercy of the fire. You can't put it out. It'd be good to get water on the tank to be sure it doesn't BLEVE, but it's difficult to get the fire out without it reigniting, with all the hot surfaces.

"It really wasn't that frightening," Tom repeats. "It was unsettling to see it catch fire, and there was some urgency to evacuate the surrounding area, but there was never any indication that the tank was going to BLEVE. There was no fire intensity change, no noise. But we didn't want to be the first ones to have a BLEVE." The fire was very low after a few hours, and they just let it burn out. The leak was small, so that took a day and a half, but there didn't appear to be any danger.

No one was injured, and there was not even any significant harm done to the truck, which was back on the road in a week. "Most of that time was spent waiting for the inspector to come," says Chris Hosford (a client of his owns the vehicle). The inner and outer tanks on the trailer were undamaged, as was the vacuum between them. An industry-wide inspection of similarly outfitted trailers ensued, and life went on.

In past accidents, fires on vehicles hauling LNG have been caused and fueled by diesel, and the LNG payload was not affected. One incident took place in Woburn, Massachusetts, in 1998. As described by the LNG-specializing consultancy firm CH-IV International in their "Safety History of International LNG Operations," the truck was traveling at high speed when it was sideswiped by a car and careened into a guardrail. The diesel fuel tanks ripped open and burned; the driver, trapped in the cab, died.

Jeff Beale, president of CH-IV, saw a report of that incident on TV. Says Jeff, "While the local press were describing the impending peril and the local fire chief was describing the 'blast zone' and contemplating evacuating everyone within a mile, LNG personnel were doing their job, righting the trailer and off-loading the LNG into another trailer."

Jeff wrote about another Massachusetts accident, in 1994, when an LNG trailer carrying more than 10,000 gallons of LNG rolled over on a traffic circle in Revere. "Traffic was diverted. The fire department and emergency response squads were called to the scene. Then the news media descended on the site. I sat glued to the 11:00 news. For the first six minutes of the newscast, the reporters painted a picture of a potential disaster of near epic proportion. They questioned the authorities' decision not to evacuate the area. They interviewed anybody they could find that was scared. Unfortunately, they did not choose to speak to anyone who understood the situation. Meanwhile, under standard procedures for this type of incident, the trailer was partially emptied into another trailer and then uprighted. But that was not the story.

"The story was, there was no story. No spill. No fire. No injuries. No disaster. As a matter of fact, neither the *Boston Globe* nor the *Boston Herald* chose to cover the story at all the next morning. When you consider the design and construction of an LNG trailer compared to, say, a gasoline tanker, it's not surprising there was no disaster."

All told, not a single member of the general public has been killed due to LNG operations since the Cleveland explosion in 1944. Not one.

Other places have been reported as sites of an LNG disaster, but the reports are at best irrelevant, exaggerated, or erroneous, and sometimes maliciously or at least knowingly left to frighten people even after their authors or the tellers of the tales know that the facts are not as described. One example is a report of "an LNG pipeline explosion" in Nigeria in August 2005 that left eleven people missing and aquatic life completely destroyed in an area of twenty-seven square kilometers. This report is posted on the anti-LNG Web site of attorney Tim Riley, as is a follow-up story that says the leak was in the main natural gas supply line into the plant.[53] This incident, like others, was tragic, but it was a natural gas event that had nothing to do with LNG except that the gas would have been turned into LNG had it reached the plant. Tim Riley knows that this is the case,[54] but he leaves the story posted on the Web site under a big red headline.

This is not the only tale to have appeared about LNG pipelines blowing up, but, at this writing, no cryogenic pipelines exist, other than within processing facilities, so there are no LNG pipelines to blow up. As in many other news stories about the hazards of LNG, a zealous reporter has simply added the "L" part to a story about natural gas, or mistaken liquefied petroleum gas, which is mostly propane—and explosive—for LNG.

Also entirely unrelated to LNG are the issues concerning the impacts of the chemical industry on a community described in "Breathing Poison: The Toxic Costs of Industries in Calcasieu Parish, Louisiana."[55] A link to this report has been posted on the anti-LNG Save Passamaquoddy Bay Web site. It would appear that a score of companies are sending tremendous quantities of toxic chemicals into the air in this modest African-American community, where rates of mortality and illness are exceptionally high. The area is home to one LNG import terminal and soon to be the site of at least one more, but these are not the cause of the toxic pollution problems. Many of the other reports with links on the site are fair analyses of LNG; why the need to post such irrelevancies?

An event northeast of Chongqing, China, in 2003 has also gotten such treatment, and has been reported in Australia as evidence that LNG would be dangerous there.[56] A natural gas explosion killed 234 people, and more than nine thousand, from twenty-eight villages, had to be treated for inhalation of poisonous fumes. What actually happened was that a natural gas well blew out, releasing highly toxic hydrogen sulfide.[57] Six individuals were put on trial for having played a part in causing or permitting the accident, but the tragedy had nothing whatever to do with liquefied natural gas.

Occasionally, one hears of a natural gas pipeline blowing up, and people who are worried about LNG then use that as an argument against importing LNG. Pipelines burst when the steel of the line has corroded. These pipelines are carefully monitored, and corrosion to that degree happens rarely. It is caused when naturally occurring contaminants—water, in particular—are carried along with the natural gas. The water is removed from LNG as it is processed, so, ironically, an argument can be made that all natural gas should be liquefied and regasified before entering a pipeline, so as to prevent that corrosion. (On the other hand, the pipelines near the Cove Point, Maryland, LNG import facility have experienced some unusual trouble with couplings that has been attributed to the lower temperature or possibly even the lack of impurities in regasified LNG.)

Undoubtedly, as time goes on, there will be more unfortunate events associated with LNG, as there will be with natural gas and with every other energy source available to us, but to date, since Cleveland in the 1940s, there has been no loss of life in the general public directly attributable to LNG. Great care is taken by all concerned to ensure that this record continues.

Questions of Cost and Supply[58]

SUPPLY and price are issues that are entirely out of the hands of hopeful developers or the operators of existing terminals. Price is dependent on supply, and vice versa, and the impact of domestic natural gas prices on LNG investments in any country is obvious.

It does cost more to liquefy, ship, and vaporize than to simply pipe unprocessed natural gas from the field. In a given market, if the price differential between domestic and foreign natural gas doesn't pay for those costs, imports of LNG are not going to happen. The first LNG boom in the United States started in the early 1970s, when LNG prices were competitive with those for domestic natural gas. Four import facilities were built in the continental United States during that decade. But then natural gas prices were deregulated during the Carter administration, deals were cut with Mexico and Canada, there were disputes with suppliers overseas about gas prices, and domestic prices dropped steeply. That was nearly the end of LNG in the United States for a number of years. Only the Distrigas terminal in Everett, Massachusetts, remained in full operation through the next fifteen years, because natural gas prices are always high in New England. The Trunkline facility in Louisiana brought in the occasional ship during the slow time; the other two plants were mothballed. Conceivably, that scenario could happen again.

It so happens that domestic natural gas is limited in the majority of the most industrialized countries, as it is in those nations which are now developing fastest. There will be competition for the imported product. The demand for natural gas, and therefore LNG, is increasing in nearly every country that now imports it, not just in the United States, but also in Asia and Europe. Import terminals are also under construction or at least under consideration in several countries not currently importing, including Canada, Mexico, Jamaica, the Bahamas, the Philippines, the UK, New Zealand, Indonesia (on West Java), the Netherlands, and Pakistan.

In addition, export terminals are being built and expanded. Russia has the largest natural gas reserves in the world and is soon going to start exporting LNG, as are Norway and Egypt. Expansion is taking place in Algeria, Australia, Nigeria, and most notably Qatar, which only began selling LNG in 1997, but whose entire economy is now based on it. New, extra-large ships are being developed for Qatar, which already has several on order. By 2012, it will be the world's largest exporter of LNG, and expects to have a 30 percent share of the world's trade.[59]

Exploration and investigation into the viability of exporting LNG are going on in Peru and Bolivia, Iran, Angola, and Equatorial Guinea. There has been a project under discussion in Yemen for a decade, although as yet no significant progress has been made. At the same time, adjustments are always being made in the estimates of natural gas reserves—from 2004 to 2005, for instance, Australia revised its estimates from 90 trillion cubic feet to 29 Tcf. (Oops.) Nigeria added 17 Tcf to its estimates in the same period, while in other places the changes were far less dramatic. But it's all a bit of a crapshoot. The local natural gas price also determines what technology is economically viable; with further increases in price, gas deposits currently uneconomical to reap become economically available and resources get larger. New technology can increase harvestable supplies, too.

A single example of the several unknowns in natural gas production is what's called "below salt" in the Gulf of Mexico. Until recently, it wasn't possible to test seismically for natural gas beneath a salt dome. The gulf may have quantities of natural gas below salt that might change the overall picture in this country significantly. Or it might not. No one knows yet. But even a temporary upward blip in domestic supplies (with a corresponding drop in prices) could put some LNG projects in financial hot water.

Political issues can have an impact on supply, as well. Indonesia, for instance, has been debating recently whether or not to renew various LNG contracts. Its requirements for domestic use are growing, and it has had trouble securing enough gas reserves near existing LNG facilities. For investors, Indonesia's system of royalties, taxation, and production sharing also make it less attractive than some other countries. For the same reasons, it is not producing as much oil as it might, either; it is a member of OPEC, but has not been able to meet its quota. It faces a general policy decision about whether or not to create a more attractive investment framework in order to widen its upstream production.

The lack of political stability in other LNG-producing nations could cause trouble, too—for instance, in Nigeria and Angola.

Then there's the quality of the product. Today, Japan imports LNG with a higher ratio of heavier, richer hydrocarbons than we do. Their import facilities and power plants are designed to use a higher heating-value gas than ours are. Our facilities require natural gas with a higher proportion of methane, and that is what our pipeline system is set up to deliver. A spot market for LNG is developing, but the variability of the product could cause physical problems at facilities equipped to handle only gas of a particular composition, which limits the sources of product for terminals that will depend on spot market purchases.

In the past, long-term contracts for LNG, twenty or twenty-five years in duration, have bound importers to particular exporters, but with the volatility in price experienced of late, exporters are no longer as enthusiastic about tying themselves down to that degree. In fact, recently a contract between Indonesia and China was renegotiated, as Indonesia felt it should get more money for its product, and the buyer in China had to agree. On another contract, for Australian gas, more aggressive Japanese buyers outbid Chinese buyers. Every day's oil and gas press reports are full of such stories.

Regardless of supply, the price of fossil fuels is volatile these days. The *Oil & Gas Journal* may post two completely contradictory stories at the same moment, as they did on November 17, 2005. One was entitled "Increased demand raises energy prices" while the other's headline read, "Warm weather undermines energy prices." At least the following day's post had the header, "Energy prices continue to seesaw."

Considerations of Terrorism

A very important question, and one given much print space and public discussion, is whether or not an LNG tanker would make a good terrorist target. An LNG carrier does represent a lot of energy in one place, and that's the reason people are frightened by it. Responsibility for the fact that they are sometimes panicked, rather than merely concerned, can be placed at the door of people such as Rhode Island's Representative Patrick Kennedy, who testified on the House floor that he'd be happy to have his gas piped in from some other country, rather than having an import terminal in Providence. "As every other state that has an LNG facility knows," he said, "if we were to ever have that explode, it would decimate a 50-mile radius."[60] Never mind that *decimate* doesn't mean what he thinks it does, no serious student of the effects of LNG spills and fires has suggested recently that there would be anything even approaching a fifty-mile radius of effect if a disastrous LNG event were to take place. (When asked for confirmation that Mr. Kennedy was quoted correctly, a spokesperson said, "I think it's fair to say the congressman was delivering a very impassioned speech and misspoke when he stated it would 'decimate a 50-mile radius.'")

It is easy to find articles and Web sites whose authors are convinced that LNG ships or facilities are tremendously attractive for violent purposes. Many refer to remarks made in 2004 by Lord Levene, chairman of Lloyd's of London, who said, "Specialists reckon that a terrorist attack on an LNG tanker would have the force of a small nuclear explosion. And it's not just the vessels, but the terminals and the whole infrastructure which are at risk from terrorism."[61]

The quotation has been used again and again as if the "reckoning" were being done by people particularly knowledgeable about the field, but the "specialists" referred to are but one: Dr. J.C.K. Daly of the School of Slavonic and East European Studies at the University of London. He is an authority on terrorism, but he's not an engineer or physical scientist of any kind. Second, the Lloyd's that Lord Levene chairs is the insurance company, not Lloyd's Register, which, like the American Bureau of Shipping, is one of several agencies charged with certifying that ships are designed and constructed in accordance with particular standards.

Lloyd's Register issued a press release a couple of days after Lord Levene made his comments, clarifying the lack of affiliation between Lloyd's of London and themselves, and making their own statement about LNG and terrorism: "Global terrorism is certainly a major threat and all reasonable measures should and must be taken to mitigate the risks and consequences of any actions, however, commentators and observers are incorrect if they believe that a terrorist attack on an LNG carrier would have the impact of a nuclear explosion." They went on to discuss the technical reasons why Daly's statement is incorrect, explaining both the design of LNG carriers and the behavior of LNG. As have others, they used the comparison with another hydrocarbon, coal: "For example, a lump of coal contains lots of energy, but when set on fire, its energy doesn't all come out instantly like a bomb. Instead, the coal burns over a period of time, releasing its energy as it goes. Similarly, LNG carriers contain large quantities of energy, but the energy can only be released slowly in the event of a spill or a fire."[62]

Lord Levene is hardly the only person to compare loaded LNG ships to nuclear bombs, however. It's an easy scare tactic. Dr. Phani Raj, who has made a career working on LNG safety, rebuts the argument in a letter to the *Providence Journal*, explaining that "the extent of damage is dependent on the rate at which energy is released rather than by the quantity of energy released." He has compiled a list comparing the energy content of various items to that of the Hiroshima bomb. An LNG ship contains the equivalent of 44 Hiroshima bombs, a gasoline tanker 58, a railroad hopper of coal has 4, and Lake Mead, confined by Hoover Dam, holds 781 Hiroshima bombs' worth of energy. However, a nuclear bomb is released in fractions of a second, he explains, while an LNG release would happen over a period of "several tens of minutes followed by a fire of equally long duration, if one occurs."[63] Still, if a carrier were to be attacked by terrorists and spill her cargo in an urban harbor, undoubtedly there would be disastrous effects.

We can no longer pretend that violent political activism has nothing to do with us. It would be foolish to ignore the possibility of a terrorist attack. September 11, as so often is said, changed everything. The Jihadists, al Qaeda, and their ilk have articulated their goals to include attacking in the United States, killing large numbers of Americans, damaging the U.S. economy and infrastructure—and, specifically, the oil and gas infrastructure.[64] Even though it's not uniquely suited for the purpose, an LNG ship that somehow could be made to spill and burn her cargo in the right place could satisfy all those ambitions.

The real question regarding terrorism is impossible to answer definitively: what is the likelihood of an attack? "Traditional risk management methodologies would have determined that the probability of terrorists employing hijacked commercial passenger aircraft to destroy the World Trade Center was zero," wrote Richard A. Clarke, former senior White House security advisor to presidents Clinton and George W. Bush.[65] You never know. And an LNG ship might be useful in various ways. Al Qaeda demonstrated the effectiveness of using airplanes as weapons—ships (LNG or otherwise) could also be utilized in that manner.

Piracy is rampant in particular places on the globe, notably off Southeast Asia and Africa, and especially in the Malacca Straits.[66] While most attacks have immediate monetary gain as their motive, it has been suggested that some incidents have been training exercises for a bigger and better hijacking, maybe using the vessel to plow into the heart of a highly populated port city. Or perhaps a fuel or chemical tanker would be attacked in order to free her cargo to do its own damage.

Fifty thousand ship movements carry a quarter of the world's trade and half of its oil each year through the Malacca Straits. Linking the South China Sea with the Indian Ocean, the 500-mile-long strait separates the Malay Peninsula and Indonesia. Bypassing the straits adds 600 miles to a voyage. It is a narrow waterway with uncounted small islands and shoals, which, along with the vast numbers of ships traveling in each direction, force vessels to travel at slow speeds. The channel closest to Singapore is only a mile and a half wide, practically a cow path for the huge ships that make passage through it. "In the Singapore Straits you might have fifty targets on your radar screen," says Captain Dick Johnston, fourteen years master on LNG ships.

This is a seriously impoverished part of the globe, an area of radical Islamist activity rife with its own national internal tensions as well as international disagreements. None of this has helped efforts to control piracy. In 2004, of 325 pirate attacks reported worldwide, ninety-three were in Indonesian waters, nine in Malaysia, and eight more in Singapore.[67]

Many of the pirates are brother-and-cousin operations, a few armed men in a small fishing vessel, but others are more heavily armed and are operating out of more sophisticated vessels, sometimes including helicopters. If they're not formally military, at least they may be recipients of military training. Often, when a ship is taken, the bandits grab the convenient stores and valuables and scurry away again. Sometimes captain and crew are held for ransom. There are occasions when a new name is painted on the vessel, a new flag is raised, and the cargo is sailed away somewhere and sold—the original ship just doesn't exist anymore. The ship may be turned back to her original crew—or those unfortunates may simply be dispatched while the phantom ship continues to operate in parts of the globe with relatively little formality about paperwork.

However, there have been times when a tanker has been taken by men armed with automatic weapons, who ran straight for the bridge without even searching for goods worth stealing or money. They simply piloted the vessel for a time, and then took off again. It is the events like this that lead some to believe that the Malacca Straits and other areas may be serving as maritime training grounds for franchise terrorists. Even if there were no effort to take a ship to distant ports to raise havoc, she could bring much disruption to the world's economy simply by blocking passage through the Malacca Straits, or Rotterdam Harbor, or the Port of Long Beach.

Most companies take every precaution they reasonably can to prevent their ships from being boarded. Going through the known dangerous zones, they may have electric fences on deck and closed-circuit cameras allowing vigilance without having to man the deck with crewmembers as sitting ducks. Instead they lock themselves inside the vessel, hoping that she will be impenetrable should anyone get aboard. They light every lamp and spotlight available, and they blast their fire monitors at anyone attempting to come alongside or climb up the sides. But few vessels carry arms; owners and unions alike believe that firearms only further endanger the lives of the crews. Most companies have made it a policy that if pirates do get aboard and into the interior of the ship, the captain should let them take what they like and be on their way. Often, the attacks are not reported, as the ship would then be held up during the resulting investigation, and would lose yet more money sitting idle.[68]

Naturally, efforts are being made to combat piracy, both within the maritime industry and by various (though not all) governments, but it's a serious problem, and one not easily beaten. Certainly the idea that a ship might be used as a weapon has to be considered. Asked in 2003 about the likelihood of an organization ramming a chemical tanker or barge loaded with explosives into San Francisco or another American port city, Rear Admiral Kevin Eldridge of the U.S. Coast Guard said, "It's likely enough for us to put a lot of effort into planning for it."[69]

According to International Maritime Organization rules, all ships above 500 gross tons now must have alarm systems in conjunction with their required automatic identification system; a push button will silently alert the nearest onshore authorities of a crisis. In this country, along with other, less public U.S. strategies, simple methods have been adopted to alert the Coast Guard on smaller vessels, too. For example, a tugboat operator can use a particular catchphrase in a routine radio security call that would seem innocuous to a hijacker but that the Coast Guard would know meant the vessel had been seized. Still, knowledge of a situation doesn't bring an end to it.

It is not inconceivable that an LNG ship would be a plausible target for pirates. In some ways it would be more difficult to attack than many other vessels; its decks are much higher off the water than an oil or chemical tankers' and would be more difficult for the average Joe pirate in a small boat to scale, but one assumes that a successful taking of such a ship would not be made by average Joes.

Causing a dramatic fire or chemical spill in any of our big port cities would accomplish some if not all of the goals that Clarke attributes to the Islamic extremists. Using an LNG tanker would not be the only means, nor even necessarily the best, but it would potentially be effective.

It's not a simple matter to significantly damage an LNG carrier, as has been well demonstrated by the accident record to date. Even running hard aground at eighteen knots didn't cause the release of any product, although, interestingly, the membrane tanker's cargo tanks were pushed out of shape in that incident.[70] But as the case of the supertanker *Limburg* demonstrates, given enough explosives (and the perpetrators' willingness to give their own lives to the job), even a modern double-hulled ship can be intentionally breached. Small boats laden with explosives did the dirty work on both the USS *Cole* and the *Limburg*. Sandia considers a number of different approaches that might be taken: scuba divers, an airplane attack, either with firepower or using the airplane itself as a weapon, or an shore-based attack using rockets or other projectiles.

The *Cole* was just a single-skinned vessel, but the *Limburg* had double hulls much like those of an LNG tanker. She suffered a huge, gaping hole in her outer hull, but damage to the inner hull was "much, much smaller," according to CDR Chris Oelschlegel of the U.S. Coast Guard, who was aboard her after the incident. Still, spewing some 90,000 barrels of oil led to an ugly, smoky fire. With her insulated tanks, an LNG carrier has one more level of protection, but it's certainly not inconceivable that one or more of those cargo tanks could be breached, particularly by serious, purposeful attack.

Occasionally, LNG supporters make remarks that are, at the least, disingenuous: "The explosive used to breach the hull would cause more damage to the surroundings than the subsequent LNG spill and pool fire," wrote one group of consultants.[71] Given the scale of a fire that could ensue, that's difficult to imagine.

One wonders if an intentional LNG release could be made from aboard the vessel, causing a vapor cloud to drift toward some target or other. According to the Coast Guard inspectors of these vessels, it would take a very knowledgeable person to accomplish all the steps required to release LNG from a tank, and there are many sensors in key areas that would set off alarms if even minute amounts of gas were to reach them. Both the Coast Guard and operators of LNG ships are making every effort to screen employees. The likelihood of the sabotage scenario is exceedingly small.

There are plenty of possible targets for terrorism in our country, targets of many types and varying vulnerabilities. There are symbolic targets, which the twin towers were, and a national monument such as the Statue of Liberty or even the Grand Canyon might be. There's always talk of the attractiveness of an event like the Super Bowl. A serious disruption in any big city would have a significant economic impact. Some conclude that flammable fuel facilities would be a less attractive target for terrorists than chemical facilities. A report made by the General Accounting Office in March 2003 states that 123 chemical facilities across the nation "have toxic 'worst-case' scenarios where more than one million could be at risk of exposure to a cloud of toxic gas if a release occurred."[72] By comparison, "Flammable chemicals affect fewer people because the distance the flammable substance travels tends to be significantly shorter."[73]

Chlorine gas, for example, is so nasty that the Coast Guard prohibits it from being transported by ship at all; only barges of a certain, smaller capacity are permitted to bring chlorine into our ports. A single rail car contains 33,000 gallons—spill a load of chlorine, and lethal concentrations could extend more than a mile, with serious health impacts being experienced for far greater distances. The EPA says that three parts per million is the maximum amount of chlorine to which most people could be exposed for an hour without suffering "irreversible or other serious health effects or symptoms which could impair an individual's ability to take protective action." Toxicity of even that level can cover an area ten miles in diameter under the right conditions.[74]

No credible modeling predicts any impact of nearly that range from LNG.

Clearly, attacking the right chemical plant would bring far more dramatic results than attacking a facility that deals with flammable substances. Look at 1984's event at the pesticide plant at Bhopal, which killed outright at the very least 3,800 people and likely more than twice that many, and perhaps as many as 20,000 total, with hundreds of thousands more still suffering from ailments caused by the accident and lingering pollution. [75]

A ship full of explosives can do a job on a harbor, as happened in Halifax, Nova Scotia, in 1917. The French ship *Mont-Blanc* was headed for home with munitions for World War I when she was struck by the Belgian relief ship *Imo*, causing an explosion that killed fifteen hundred people immediately and hundreds more in subsequent days. Nine thousand people were injured by flying building debris, and shrapnel from the vessel herself landed five kilometers (three miles) from the harbor. There was nothing left on the shore within nearly two kilometers (a mile and a quarter) of the ship. As if that weren't enough, the blast caused a tsunami ten meters (thirty-three feet) above high water on the Halifax side of the harbor.[76] Many regulations are now in place regarding shipping hazardous materials, and one would hope the Halifax explosion couldn't be repeated, but the event does show what's possible, even accidentally. A knowledgeable and purposeful terrorist could presumably be as effective.

More than forty-eight million tons of hazardous materials such as LPG, anhydrous ammonia, and chlorine were shipped through U.S. waterways in 2003, with much of that cargo going to and from urban areas.[77] Truly, LNG offers little more to a terror-intent individual or group than already exists here. That's not to say that LNG carriers, processors, and regulators should ignore the threat of terrorists.

Listening to Stakeholders:
The Licensing Process

IT is because of the recent flurry of new LNG facility applications, and rumors thereof, that many people have become aware of LNG at all. The application process is long and involved, and gives people plenty of time to learn about the issues facing any particular facility—and plenty of time to get upset. Those issues come down to the possible impact on the local physical and human environment.

When the federal government enacted the National Environmental Policy Act of 1969 (NEPA), it accepted responsibility, at least to some degree, for overseeing the impact of man's activity on the environment. NEPA has been amended at various times since, but it states its purpose in part: "To declare a national policy which will encourage productive and enjoyable harmony between man and his environment; to promote efforts which will prevent or eliminate damage to the environment and biosphere and stimulate the health and welfare of man; to enrich the understanding of the ecological systems and natural resources important to the Nation."

NEPA recognizes the profundity of the impact of our activity on our world, and continues: "The Congress authorizes and directs that, to the fullest extent possible, the policies, regulations, and public laws of the United States shall be interpreted and administered in accordance with the policies set forth in this Act." It then directs all agencies of the federal government to concretely consider the environmental impact of any proposal.

In the case of LNG terminal siting, FERC has the major task of creating the Environmental Impact Statement (EIS); this document is the basis for approval or denial of the license. They draw upon information provided by the applicant and on independent third-party consultants to actually draft the document.

Almost all, if not all of the fifty states have also created their own state environmental policy acts to further ensure comprehensive environmental reviews of almost any sort of significant developments, energy-related or otherwise. Sometimes, the federal and state requirements are so different that developers must create two entirely separate sets of studies.

Onshore

The process of getting a permit to construct a shoreside LNG facility is certainly complicated. The way is tangled and convoluted, with many issues to be addressed, seemingly infinite numbers of permits to be obtained from overlapping jurisdictions (perhaps as many as fifty), and at least one comprehensive Environmental Impact Statement to be created and reviewed by various agencies. But the biggest hurdle is the Federal Energy Regulatory Commission, where applicants must deal first with the staff and finally with the commissioners themselves—five or three or however many happen to be sitting when the time comes up. The Federal Energy Act of 2005 gave FERC the position of lead agency for licensing onshore LNG terminals. An individual state may appoint a state agency to represent its interests with FERC, but to FERC and FERC alone is given the ultimate authority to decide where and if an onshore LNG terminal may be constructed. "The Commission shall have the exclusive authority to approve or deny an application for the siting, construction, expansion, or operation of an LNG terminal," the act states clearly. (California and perhaps other states are challenging this pronouncement in court.)

FERC maintains an excellent Web site—www.ferc.gov—with links to good information about each of the industries it's involved with. Most important for our purposes here, there is not only an up-to-date library of the record regarding every LNG project or application, but they also maintain a list of predefined searches for the records of existing and proposed terminals. Any time any paperwork is filed with or created by FERC, be it required information from operator or applicant, request for information from FERC, response, rebuttal, comment or decision, it is posted within the docket file for that project. Throughout the rest of this book, there will be references to these dockets and their contents. They're not entirely easy reading, but the information is all there. That is, it's all there except for any illustrations or tables, which for some reason are only available in person, and anything that has been deemed sensitive from a security standpoint.

Listening to Stakeholders: The Licensing Process

The process of getting a shoreside project approved in the United States has a number of chicken-and-egg aspects despite (or perhaps because of) FERC's best efforts to streamline the process while sorting out serious and viable applicants from wannabes. (There have, in the past, been a number of proposals floated by people who don't have the backing or experience to carry off such an immense scheme. Under the older FERC rules, they wasted a great deal of the government agency's time and resources, as well as their own, before folding their cards, which ultimately benefited no one.)

In an attempt to speed up the application and review process, FERC now requires a pre-application phase designed to hear the concerns of "stakeholders"—people who might have an interest in the project, including state and local officials; community, business, and environmental associations; law enforcement and fire people, and anyone with any concern whatsoever. Environmental and other issues thus will be identified early on and can be addressed during the rest of the process. In the past, applications sometimes were filed without people who might be affected having any knowledge of the project. Pre-application hearings are helpful to FERC staff, as well, alerting them also to the issues involved. A final application may not be filed until at least 180 days after the pre-application process has been approved by FERC.

The idea of the pre-application was undoubtedly a good one, but the process isn't so simple. It's tricky for a project developer to work his way through the traps that await him, starting with the timing of his request to be accepted into a pre-application phase. Here's the first chicken-and-egg situation. A sensible developer will have already identified people who will have concerns about the project, and spent time working with them. How much more effort should go into analysis at this time? Certainly a fatal-flaw analysis is necessary to be sure there's no immediately identifiable obstacle that will preclude completion of the project—there's adequate water access, there are no legal barriers to gaining title to the needed property, and no endangered-species habitat is affected, as examples.

The pre-application phase is the first of a number of levels of study, and has to be done before going fully public or involving any permitting agency. But there's no sense in spending a lot of money on full studies that may then be deemed inadequate or contrary to FERC's wishes; two or three million dollars might be spent on an engineering design that is entirely dismissed by the reviewers when the time comes. Yet the more information the developer has in hand, the better able he will be to answer questions from the public (who naturally enough want answers as soon as they can get them) and the better prepared he will be for the ultimate requirements of his application.

FERC says it likes to see local support for a project. In rural places like Maine, town government may have the opportunity and grounds to vote on the suitability of a project (and, in fact, two Maine towns had turned down LNG projects and residents of a third one had made it perfectly clear they weren't interested before Robbinston finally voted in favor of one). A good question, and a potential dilemma, is, should a developer wait until after a town's vote on his project to make his request of FERC, running the risk of appearing insincere if the town vote is against the project, or should he go ahead and make his request of FERC before the town's vote, thus showing progress and credibility even though he might risk being accused locally of not caring what the town thinks?

From a small municipality's point of view, it's not simple either. Does it want to have its say, whether by a popular straw vote (as has taken place in Maine towns) or through more formal channels such as planning board approval, before all the research and reporting is done? When volunteer members sign on to their rural planning board, they are unlikely to expect such a large-scale, high-visibility, high-value, and complicated application to come before them. Often a multi-unit subdivision is the biggest project they've been required to address. They cannot be expected to already have significant knowledge of the complex issues involved should an LNG project turn up. Still, a developer doesn't want to spend millions of dollars on reports, and then be turned down by the local community.

A question with contradictory answers is whether a local community can, in fact, reject the siting of a project based on its own regulations alone. Even within FERC, disagreements on this subject seem to exist. (More on this in later chapters.)

One advantage the LNG developer has over the smaller-scale real estate developers most cities and towns usually deal with is that, should his project go through, both he and the community stand to do very well financially. An LNG facility developer can make economic promises that most other developers can't. If the local fire or police department is going to need new equipment or more personnel because of his project, or a road needs to be upgraded to handle extra traffic during the construction phase, he can offer to pay for it. If fishermen's gear is going to be damaged, he can promise to replace it. If a school building is deemed too close to the project, he can build a new one. These offers may be taken with skepticism in many areas—what shipping company has ever suggested replacing a lobsterman's gear before? But sincere developers will make their promises the subject of contracts that will be binding on any future owners or operators of the facility in question.

A couple of formal requirements need to be met before making the actual pre-application filing (is that a pre-pre-filing?). The developer must first write and have FERC approve a Request for Proposal to find consultants interested in creating the third-party NEPA-required Environmental Impact Statement. From the responses the developer must come up with at least three acceptable companies, from which FERC selects one. (Interestingly, sometimes only a single FERC-affirmable response has been received, leaving one to wonder both about the difficulties of working for FERC and the quality of work the sole respondent is likely to provide.) Also prior to the filing, the developer must send a Letter of Intent to the Coast Guard, accompanied by his own preliminary Waterway Suitability Assessment (WSA). (We'll come back to that, too, later in this chapter.) Only then can he request pre-application status from FERC.

The paperwork accompanying the request for pre-application review may only be a few pages long, but must include a brief description of the project and what work has been done already, plus a list of interested parties, public and private, and their contact information. These criteria must be met in order to set up the scoping meeting that FERC requires after accepting a filing.

During this pre-application time, the project manager, the person heading up the application process, is extremely busy identifying and acknowledging problems and solving them. He is making sure that the thirteen specific resource reports are generated that will describe not only the project but the existing and potential environmental impacts, many and specified; drafts of these must be handed in to FERC at least ninety days prior to filing the formal application. He is also arranging for the preparation of engineering and design documents for the final application. All materials must be given to FERC at prescribed times. (And never mind that the regulations and guidelines effectively preclude meeting that timeline. "Too bad," one applicant was told, when he remarked on that fact to someone at FERC.) The project manager also is ensuring that state environmental requirements are "trued up" with similar or sometimes contradictory federal requirements. In addition, through this entire process, he makes himself available to visit the site and attend other meetings arranged by FERC's staff, while simultaneously filing monthly status reports with FERC. He is arranging for a "follow-on," a more complete Waterway Suitability Assessment that must be filed with his final application. And, if all the funding needed to go on with his project is not already lined up, he and the developer are looking for that as well.

Another chicken-and-egg situation, one that's legal in nature, needs to be addressed during the pre-application phase: one must gather approvals and sign-offs from a number of state agencies. In Maine, for example, they are the State Planning Office, the Historic Preservation Commission, the Department of Conservation, the Department of Environmental Protection (with ten licenses, permits, and certificates), the Department of Transportation, the Fire Marshal's office, and so on. FERC says that the state and local permits should be gathered at this time—but without the formal authorizing that a filed FERC application provides, the state agencies can't consider requests for permits before them. The project manager can easily feel as if he's in a Catch-22. In some cases, depending perhaps on who knows whom and which politicos have an interest, state agencies might do all the preliminary work on their various approvals. Then, when the final application is filed with FERC, all they have to do is issue their particular permit (or hold whatever public hearing is required and then issue it). However, the applicant can't count on this happening.

But lest anyone get the wrong idea, while there may be a number of state and local agencies that get to check out a proposal, they don't necessarily have the power to prevent the project from going forward. Even though former FERC commissioner Nora Brownell says, "If people really, really don't want a project, there are ways to stop it," and developers fear that to be the case, it may not necessarily be so, as has been demonstrated in some municipalities where an application has been made and considered favorably despite clearly stated opposition from local officials. More on this in subsequent chapters.

"States may void an import or export project through imposing federal authority delegated to them under the U.S. Coastal Zone Management Act, the Clean Air Act, or the Clean Water Act," wrote Tamara Young-Allen, spokesperson for FERC on LNG matters. But, she says, states and communities may not rely on their own laws and standards to stop a project; they do not have a veto right that impedes interstate commerce, which Congress has defined specifically to include LNG terminal placements. They may possibly have some say about just how the project will be completed.

Eventually, the developer files his formal application. The selected consultants go to work on the Draft Environmental Impact Statement (DEIS) and the applicant turns his attention to the myriad other requirements, such as the use permit reviews and approvals from the federal Fish and Wildlife Service, the National Marine Fisheries Service, and the U.S. Department of Transportation, which oversees all the engineering design work.

And at any time during the process, FERC can decide that it wants another study completed, perhaps marine studies over an entire year to consider seasonal species, or further detailed geologic borings to ensure adequate seismic analyses, or additional soil samples to document the absence or significance of on-site pollutants, and even complex ship modeling to document in advance the LNG ship maneuverability along a specific waterway. The financial burden of these additional studies is an unknown and uncontrollable factor for the developer, and one that has been known to throw a schedule totally askew. It's a wonder that any project managers have any hair left.

No issues are supposed to be swept under any rugs during the preapplication process, but a time for public review and comment is built into the system after the draft EIS is issued. Perhaps inevitably, some individuals feel that the study missed significant issues, that it was inconsistent within itself, that it didn't really consider alternatives at all, or even that the whole thing was a put-up deal by those who want the project to go through, which the opponents sometimes claim includes FERC. A two-month comment period after the draft is issued allows for responses. Each of the many comments received must be answered by FERC before the final Environmental Impact Statement is issued. Despite a supposed effort to speed up the whole process, no time limit is placed on FERC to do this. Remarks Bob Arvedlund, retired chief of FERC's Environmental Review and Compliance Branch, "You're not allowed to just ignore what you'd like to ignore. Often people say 'You missed this or that,' and we say, 'Oh, we know about those; we didn't think they were significant.' Or we add a paragraph or two." Sometimes, people feel that not much attention is paid to their arguments. "If it came in the door," says Arvedlund, "someone's gonna respond. You may not like the answer. There's always somebody doesn't want it."

"It's very easy to enter the FERC playing grounds," says Rob Wyatt of Downeast LNG. "It's very expensive to stay there." Most likely, by the time he reaches draft permit issuance, a developer will have spent a minimum of seven and perhaps as many as ten million dollars. That's just the beginning. There will be a slug of further permits to be obtained—dredge and fill permits from the Army Corps of Engineers (even if there's no dredging or filling, a pier comes under their jurisdiction), a water quality certificate, if needed, and clearance from the state's designated coastal zone management agency. The approval may have been granted subject to any number of other specific requirements that must be fulfilled before construction can actually start. And the final, site-specific construction engineering plans must be prepared for bidding—another chicken-and-egg situation, for is it better to spend the money for what is termed the Final End Engineering Design plans during the period before the draft EIS is issued, or is it better to wait for a favorable EIS and then take a great deal more time later? The decision involves millions of dollars and months of preparation work.

The engineering is covered by two main sets of regulations. The governing document for all applications, onshore or deepwater, is in the Code of Federal Regulations, 49 CFR part 193, Liquefied Natural Gas Facilities: Federal Safety Standards. This gives the general requirements. Specifics for onshore projects are spelled out by the National Fire Protection Association, in their NFPA 59A, Standard for the Production, Storage, and Handling of Liquefied Natural Gas, which is updated every five years or so. Says Arthur Ransome of CH-IV, "The regulations are a cookbook for the engineering design contractor." Siting is a big issue, and much of the detail in NFPA 59A relates to the layout of equipment and the potential for spillage from pipes or a tank, both for resulting fires or flammable vapor. These issues are significant and the subject of some controversy, which will be addressed in a later chapter.

Details of construction are specified, too. There are details on simple things such as the tightening of bolts on flanges: too loose, and a leak might occur; too tight and the bolt might fail with temperature changes, so a special tool measures the torque applied to each bolt to ensure that it's done as specified. And welding—it is critical that every weld on a stainless steel unloading line is correct, for example, so each is to be x-rayed, and provisions in the code specify the quality of an acceptable weld. A record will be written of each X-ray, thousands of them in a newly constructed facility, and that record is to be kept, along with tens of thousands of other pieces of documentation of quality control. At Eco-Eléctrica, Arthur says, they had a library with several hundred volumes of this paperwork, along with all the equipment manuals, and so forth.

While the usual bureaucrats ascertain that every single piece of paper required has been received at FERC, FERC hires independent consultants to review all the engineering work in the same manner that the EIS consultants are hired. The idea is to identify any problems along the way and rectify them before construction, but even that's not the end of FERC's involvement. FERC monitors the construction and, finally, the operation of any LNG terminal; for the rest of the life of the facility, it will be subject to semiannual audit and biannual inspection. All modifications must be documented in the required, plant-specific Cryogenic Design and Inspection Manual, the permanent record of the history, design, and status of the facility.

Interestingly, in Europe the approach to permitting is entirely different. In this country the LNG regulations are prescriptive, spelling out just how each feature shall be built, but in Europe the developer does a risk and probability study. He can design his plant however he likes; if he can show that his plant's safety is such that no more than a specified number of deaths can be expected in a million years, then he automatically gets his permits. Governments have no discretion the way they have here. Developers use data from many industries, including other petroleum businesses, refineries, and so on, to calculate their risks.

Arthur Ransome grew up in the risk-assessment system in the UK, and is very comfortable with it. "You do your thinking up front. You do your engineering design, you determine how to construct and operate it, and at every stage, you have to do risk assessment. It's a lot more time consuming, but once it's complete, you don't have to change anything." If you're building a brand-new project, you are inventing the wheel, but next time you can use the information available from the first as the basis for conducting your risk assessment. Arthur admits to being frustrated sometimes by the American system. "I think we'd benefit from risk analysis." He speaks of a plant designed beautifully on paper, but someone forgot to figure in that the facility would actually be operated and maintained by people—their needs should influence the design. "I'm an operator," he gives as an example. "I have to isolate a pump. Is it better for it to have an isolation valve, or be welded, or have a flange? How do I get it out? Do I have to remove the roof and walls? Do I need a crane?" You have to think of how you're actually going to operate each aspect of the facility.

FERC has now added a "hazard operability" aspect to the analysis. *Hazop* is the buzzword. An independent engineering firm looks at every single piece of the plant, with a team of people, including design engineers, operators, and owners with their engineers, asking questions about operating and maintaining. The engineers decide ahead of time how changes will be incorporated. In the UK, this came about as a result of a huge accident in a chemical plant where a number of people were killed. One finding of the investigators, after the fact, was that there had been inadequate consideration of the effect of a change on other aspects of the operation.

"You put the two together, the cookbook and the risk assessment," Arthur says, "and you will have done a lot more work to understand the design of a project."

Frank Katulak of Distrigas describes the Far East as wholly different from either the United States or Europe. They've not done any fire or spill studies, although they have referred to the European ones. "They work on an entirely different philosophy," he says, "belt and suspenders." They have large and numerous LNG facilities—in Japan, there are twenty-five terminals altogether, and in Tokyo alone, there are five. "One is fifteen times the size of Everett. They're shipping every ten hours. Their storage tanks are underground, and there are buildings above them. Hundreds of people work in the plants," says Frank. "We have one day's reserve here— there, they have a month's reserve. They're unconcerned with the effects of a spill." They build their facilities as safe as they possibly can—that's the belt and suspenders thing—"and then they don't think about it. They have no alternative."

Considering an Application

For all the expense and difficulty of getting through the process, as of this writing, only one LNG import terminal application has been turned down by FERC, although others have disappeared for other reasons. During the rigamarole, differences between FERC and an applicant get hammered out. No doubt, in many cases, this happened at great expense to the applicant, but to date, save in Providence, approval was ultimately granted, albeit perhaps subject to prohibitively onerous conditions. Certainly there will be more approvals than completed LNG projects. Many final decisions are appealed in court; Providence's KeySpan turndown was appealed (unsuccessfully), and in other cases, an approval may be appealed by an intervenor who believes the approval was granted in error.

Nora Mead Brownell had been on the Federal Energy Regulatory Commission since 2001 and was within a few months of stepping down when we spoke. She came to the position from the Pennsylvania Public Utility Commission, where her particular interests were in electric choice, known in other states as deregulation, and telecommunications. That she served as the president of the National Association of Regulatory Utility Commissioners shows that she was respected in the field, despite a previous history with no ties to utilities or power at all. Her other political experience came a number of years earlier, when she was deputy executive assistant to Pennsylvania's governor Richard Thornburgh.

Among her duties in the governor's office was finding suitable candidates for the various state commissions. "We hated independent commissions," she says. They'd be appointed, and then they'd go off and do as they pleased, not what the governor wanted at all. But it is that freedom that she appreciated as a member of such an independent commission. "We have the luxury of doing the right thing, the best we can," she said. "You can't get involved politically." Because FERC is not allowed to have more than three members from either party, it gives the impression that the decisions are political. "But it doesn't come down to R and D, but rather how you see the record," she said. "Because of the jurisdiction and the mandate, if you tried to play politics and ignore the record, you'd be slammed by the court. Periodically we do get reprimanded by the court, if the record doesn't support our conclusion."

It turned out to be both an advantage and a disadvantage having only three members on the commission, as was the case for the last part of Brownell's term. "With three, it has to be a functional three," she said. Some commissions have a history of dysfunction. But it's easier to get a coherent decision from three than five, she said, "particularly with very distinct personalities. It's awkward with a two-to-one decision. Three-to-two is easier. It's a challenge to never personalize it."

Some chores and decisions are very clear—pipeline siting, for instance. It might seem to be particularly difficult because there are so many individual landowners to deal with, but it actually is a relatively easy job, she said. The staff has a great deal of experience with pipeline siting, and has developed the skills to work everything out before it comes to the commissioners. But most of the issues they face are complex, and for making new rules or restructuring an entire industry, a full team of five commissioners is better—they come up with more ideas. An example that they were struggling with at the time of our conversation was the reform of open-access electrical transmission, which dealt with engineering, economics, how markets work, and issues of equity, all of which had to be balanced. By contrast, LNG, where the issue is safety and the solutions are engineered, the analysis is more technical and the answers are generally clearer, she said. The "cookbook" that Arthur Ransome mentioned earlier also works for the commissioners.

The Coast Guard's Role in All of This

The Coast Guard plays a number of important roles in the world of LNG importation. The most obvious one is as provider of safety and security, just as when any ship enters an American port, although they are more active in supervising LNG than most other cargoes. They play a part in the planning of any new terminal, too, whether onshore or off.

Alan Moore, retired from the Coast Guard with the rank of captain, has come back on board as a civilian to head up the Coast Guard's ongoing review of multiple proposed LNG projects in Maine. "I missed the camaraderie," he says. "I don't have to carry a pager, and I can work regular hours. I don't mind sitting back and letting someone else be responsible." His background suits him for the job, starting with his involvement in assuring the safety of LPG ships coming into Portsmouth, New Hampshire, twenty years ago.

Applicants must file a Letter of Intent and a preliminary Waterway Suitability Assessment with the Coast Guard prior to filing their pre-application with FERC. The WSAs consider the nature of the port itself—its physical layout, placement of populated areas, critical infrastructure and key assets, and what impact the proposed facility would have on those things. Both the LNG facility and the tanker route to it must be addressed in detail, along with any hazards to be faced along the way.

The talk about a project predates any filings, however, and, as Alan says, "Once the letter comes in, everything gets rolling fast." A few months before any formal paperwork began on the three Maine project proposals, he was already down in Washington County checking out the scene. He went by land and water to each of the sites being discussed. He met with the pilots, who have guided equally large ships in and out of the bay in the past (without the assistance of the large tractor tugs that would be available for the LNG ships), and they assured him that they're not alarmed at the prospect of piloting the carriers. He made note of the obvious issues like the tides and currents, the local fishing, the pilots' practices, and environmental matters. He saw what support structure was in place and what wasn't, such as fire and public safety protection. Because the ships would have to pass through Canadian waters, he checked to see whether there were already established cooperation agreements with Canadian pilots (which, in fact, there were not; the two pilotage groups have worked cooperatively forever with no formal documentation).

The question arose early whether the vessels headed to U.S. ports even had the right to pass through Head Harbour Passage. Members of the New Brunswick community of St. Andrews, on the other side of Passamaquoddy Bay, objected to having such large vessels traversing the narrow, twisty passage. The "right of innocent passage" refers to the right a foreign vessel has by international convention to navigate expeditiously through one nation's territorial sea on her way to another country's port facility. The Coast Guard asked the State Department to interpret this law of the sea, and was assured that Canada would not be able to interfere with vessels passing through Head Harbour Passage to LNG facilities in Passamaquoddy Bay.[78]

Of course, the Canadian government may not see the case in the same way, and there is a vociferous group in New Brunswick hoping for just that. An amusing situation arose when the local parliamentary representatives gave the prime minister an ultimatum about making a statement on the subject. One wonders what the "or else" aspect of the challenge was. The question has not been resolved, either, as of this writing, many months after it was raised.

The next job for Alan and his fellows in the Coast Guard's Sector Northern New England is to review the preliminary WSAs, following the guidelines set out in the Navigation and Vessel Inspection Circular (NAVIC) No. 05-05, produced specifically to help the officers faced with the task of reviewing the water access to proposed LNG import facilities. It is entitled "Guidance on Assessing the Suitability of a Waterway for LNG Marine Traffic," which one might assume means that Alan and his team will come to a conclusion about whether or not Head Harbour Passage and Passamaquoddy Bay are well suited to the transits in question, but that's not quite the case.

Under Coast Guard regulation, and with their oversight and control, ships huge and small pass through the nation's 361 ocean, lake, and river ports. Everywhere, hazardous material of one sort or another is delivered and shipped out. All hazardous cargoes come under the Coast Guard's careful eye, both as to safety and to security, a distinction that once wouldn't have meant much to most of us, but today, post-9/11, security is a bigger issue. The Coast Guard can and will deny permission to a particular transit if a company or vessel can't meet its requirements.

"The Coast Guard's role is not to give a yes/no answer to an LNG facility, however," says Captain Mary Landry, Chief, Maritime Safety Division, First District, U.S. Coast Guard. Her area covers the five coastal zones along the East Coast from Canada through New Jersey. In her earlier position as commander of the Marine Safety Office and captain of the port in Providence, she was directly involved with LNG import terminal applications in Providence and Fall River. (She also was the federal on-scene coordinator for the cleanup of Buzzards Bay after the big oil spill of April 2003, but that's another story.)

More importantly for the matter at hand, Captain Landry has also been involved with Coast Guard LNG oversight overseas. U.S. Customs has established the practice of sending teams to foreign ports, particularly container ports, to be certain that they, too, have effective security programs. "We call it stretching the borders out," she says, while admitting that the countries involved "might see it as acts against sovereignty." In most cases, however, the shipping companies welcome the oversight on loading because it will allow them to unload more quickly on the American side, which, in turn, allows them a quicker turnaround for more pay dirt. They pick up the tab for the airline fare for U.S. Customs officials to come over and inspect as they're loading. "Of course, you have to make sure nothing happens in between," she says. Coast Guard personnel also have traveled to the LNG export plant in Trinidad from time to time and have ridden the ship back.

The Maritime Transportation Security Act of 2002 (MTSA), enacted in response to 9/11, recognized the ports as spots potentially vulnerable to terrorist attack. While focusing on the sorts of activity that would have a greater likelihood of being involved in a security-related incident, the MTSA spells out security requirements of all vessels and port facilities (including inspections and security plans) and establishes a formal structure of responsibility within each port and district. An international equivalent also exists, the International Ship and Port Facility Security Code, on which the United States and 147 other countries collaborated. The International Port Security Program identifies foreign countries that might pose a security risk to international marine transportation, and the Coast Guard works with that program as well.

Even prior to the MTSA enactment, the Coast Guard had been boarding "key" vessels as they transited in and out of port "to keep those vessels from being used by terrorists as weapons."[79] The requirements now are much increased. Ships must give 96-hour notice before entering an American port, providing information on the crew and the vessel, which is all vetted prior to their entry. The Coast Guard continues to board and inspect LNG ships as they approach (as they do various others for both safety and security reasons). When Captain Landry was in Houston, she used to like boarding the LPG ships. "Those ships, just like the LNG ships, were immaculate and very, very safety-conscious. With the old tankers carrying tallow or other lower-value cargoes, you had to be on your toes. Dog tankers, we used to call them, and you knew it would take a lot longer to inspect one."

After September 11, Boston's LNG carriers came under much scrutiny and, in fact, were stopped from entering the harbor for several weeks. Although the ships have always been escorted in and out, security was beefed up when the port was reopened to the ships. Since then, the Boston protocol and level of protection has come to be seen as a standard. When the plan for a particular project is described in an EIS, the published wording simply states that there is a robust security plan on the order of what is done in Boston. Of course FERC has a copy of the complete plan, with extensive requirements according to what's appropriate for the particular waterway, but it's not seen as a good idea to announce to everyone just what the plan involves, so it is not freely distributed to the public the way most of the other paperwork is.

It's a given, however, that the Coast Guard requirements will be a strain on local resources. Included in the MTSA was $170 million to cover grants for port security; Boston took advantage of that funding to boost up its marine patrol and other resources. Distrigas now pays various local and state entities $20,000 toward security costs each time a delivery is made.

But other harbors may not have the infrastructure to begin with. A town in down east Maine, for example, has no helicopters or even harbor patrol boats or the manpower for such things. The Energy Act of 2005 established that the company receiving the LNG will reimburse state and local security costs and also pay to increase the resources as needed. "It just gets built into the cost of doing business," says Captain Landry. "Normally, we don't get into that," she says, but sometimes, it makes sense. The Coast Guard can shift manpower around from one place to another when there's a crunch—for instance, a Maritime Safety and Security Team from Boston gets sent to Portsmouth when an LPG ship comes in, perhaps ten times a year—but local officials don't have access to more personnel when they need it.

"Our job is to lay out the safety and security issues to be addressed by the applicant," says Captain Landry about the Coast Guard's role in shoreside LNG terminal placement. They must determine how to make passages and landings safe for all concerned and as secure as they can be. In the case of onshore LNG terminals, it's up to the individual project's bean counters to determine whether the value of the product outweighs the extensive precautions the Coast Guard might require in a particular location.

Inevitably, some of the specifics of an Incident Action Plan must be kept quiet, and are marked SSI, Sensitive Security Information.

Offshore

The Deepwater Port Act, first passed in 1974, established a licensing system for various port structures outside the U.S. territorial sea, including LNG terminals. (The territorial sea extends twelve miles out.) It gave the Department of Transportation authority to license these facilities, and in the case of LNG import terminals, handed the application process off to be administered jointly by the Maritime Administration and the Coast Guard. Among the tasks for the Coast Guard is overseeing the development of the environmental impact statements required by NEPA, along with the review and approval of the engineering and operations aspects of the project. In effect, the Coast Guard holds the position parallel to the FERC staff's in onshore applications, while MARAD fills a role equivalent to that of the Federal Energy Regulatory Commissioners, making the final determination of whether to license a new facility or not. MARAD staff have to review other aspects of any proposal, particularly the financial viability of potential licensees and their citizenship. They also receive input from many other governmental departments—a dozen or more, from the National Oceanic and Atmospheric Administration and the Department of Defense to the EPA and Department of State—as well as the Coast Guard, but overall, the process is very similar to that of onshore applications, although some of the specific major issues are different.

For example, in an offshore terminal review, MARAD and the Coast Guard must consider the marine mammals and fish that make their homes there or that migrate through annually, as well as the fishermen who depend on having a healthy fishery. There are issues not only of encroachment on fishing and breeding grounds or habitats; they must also consider how the technology of regasifying might harm the ecosystem of the area. A major concern of environmentalists and fishermen alike has been the impact of a terminal's use of water on the feeding and reproduction of the local marine life. As seawater passes through a processing system, the larvae and plankton and small creatures it contains are killed; obviously the less water used, the less impact. Most of these worries have been allayed by the applicants' willingness to use a closed-loop system, and recent technological improvements to such systems have lowered the need for water even more.

In addition to fishing, other potential issues facing proposed deepwater LNG ports are navigation and mineral extraction. It would make no sense to site an LNG import facility in the middle of a high-yielding domestic oil- and gas-producing area, so the Federal Minerals Management Service must weigh in on any application.

The offshore pipelines also require a different set of reviews than are needed for onshore facilities.

The Coast Guard's role in approving offshore terminals, just as the FERC staff's with onshore facilities, is to identify potential impacts of a project, not to say aye or nay. Ultimately MARAD, having weighed all the information, makes the determination.

Standards

THE standards by which applications for LNG projects are reviewed are all specified in federal law—or that's the idea, anyway. The two primary codes are the Department of Transportation's 49 CFR 193, LNG Facilities: Federal Safety Standards, which in turn refers to NFPA 59A, the National Fire Protection Association's Standard for the Production, Storage, and Handling of Liquefied Natural Gas. Along the way, FERC sometimes creates written guidelines intended to help applicants comply, and sometimes there seem to be guidelines that are not written at all. Officially, however, those two codes govern everything.

Let's look at the standards themselves. There are two issues: are the intentions of the standards correct, and do the standards say what they are meant to say? We'll look at the latter first.

Well-intentioned congressional legislation turns into ever-increasing pages and volumes of regulations augmented by infinite numbers of guidelines that the well-advised applicant will follow. Talking with engineers such as Neal Frangesh, who has served on committees charged with helping to draft regulations, one understands how this happens. "They're excruciatingly painful reading," he says of regulations in general, but they needn't be. The engineers write rules that make sense to them and that they believe are easy to read. Then those rules are codified by government lawyers. "They'd take intelligent, cogent paragraphs and break them up into three or four pieces, and in the process, the meaning would get distorted. You're not allowed to have two thoughts in any one paragraph."

Neal's partner in LGA Engineering, George Randall, points out another hazard. At times, the government has been concerned about regulations being drafted by people "tainted" by contact with the industry being regulated. That's a legitimate concern—but it's also important to have regulations written by people who know what they're talking about.

Regulations evolve through time, and sometimes processes change along the way, too. More than once I've heard the expression "Make it up as you go" in reference to FERC and its requirements, both from developers and from people opposing development. I'm told that when Bob Arvedlund held the helm of FERC's LNG program, his ship was tightly run but not always predictable. New demands would be expected of an applicant, demands for information beyond what the regulations called for, but that Bob felt were important. From working on the EcoEléctrica project, Rob Wyatt remembers a particularly troublesome pre-filing request from Arvedlund—the demand to see a gas supply contract. "But we ended up getting out of that one, at least prior to our application—there was no way we could have gotten a gas contract done before the permit application." No one upstream will spend time with a project until it's real.

And now that Bob Arvedlund has retired, in today's rush to develop import terminals, many more staffers, by necessity, are working with LNG facility permit applications. Perhaps there's a less dictatorial system of management now, too, which may also mean even less clear directives. From the applicant's point of view, the situation is the same, though, with increasing and unpredictable demands. This evolving process causes hardships both for developers and for those who would like to see those developments either go away altogether or go somewhere else.

Many years ago, flying out of Gillette, Wyoming, Rob Wyatt found himself sitting next to senator Alan Simpson. They got talking. The senator asked Rob what he did for a living. "I write mine permit applications," Rob said.

Simpson thought he was kidding. He wasn't. "But what do you mean, you write mine permit applications?"

Rob explained that his job was to wade through the morass of paperwork required of a company that wanted to get a permit for a coal mine approved by the Office of Surface Mining. He had just finished making eighty copies of one application, each filling thirteen three-inch-thick binders, and including such details as how big the lettering would be on the road signs they had to post.

The senator was incredulous. He said that he expected the coal mine permit bill he had sponsored to require thirty or forty pages to be filed, a hundred at most.

Third-party contractors (TPCs) are in a difficult position, or at any rate, they're perceived to be so. The NEPA-required Environmental Impact Statement (EIS) is created by such a TPC, paid for by the applicant but working for and under the oversight of FERC. Some people take exception to the qualifications put out by Natural Resource Group, the TPC working on Sound Energy Solutions' Long Beach application, in which they proclaim their expertise to be in "securing environmental permits." Natural Resource Group claims they are "the only nationwide firm solely dedicated to environmental permitting and services that advance the construction and operation of pipelines, storage and LNG facilities." But shouldn't a third-party analyst working for FERC be looking objectively at the situation before him and delving into its pros and cons, rather than automatically taking a stance in favor of the proposed development? The client is not the applicant, after all, but rather the government agency that is charged with protecting the citizenry. And yet, say some, EISs invariably conclude that the project under consideration is a fine one. (From the applicant's point of view, however, a positive stance by the TPC cannot be assumed. It's all a matter of perspective.)

Given that FERC starts with the assumption that the country needs LNG, its own job is to be sure that LNG reaches our pipelines safely. Reaching an agreement on a plan is just what the consulting team is meant to do. Any and all issues facing a project are supposed to have been identified, and solutions or mitigation devised, by the time the final EIS is completed. In theory, everyone with legitimate concerns will have received satisfactory answers, and everyone will be in agreement about the project. The difficulty with that thinking, of course, is that judgment must be made about the legitimacy of concerns, and not everyone feels that their views are being considered fairly.

Perhaps FERC's job description should be slightly different, however, and that's where questioning the standards comes in. Still assuming that the country needs more LNG, many people feel that the task before FERC should be to assure that the United States gets the product in the safest manner possible. It doesn't seem to be that huge a discrepancy between the terms "safely" and "as safely as possible," but the implications are vast.

Ever since the 1970s, again and again, one entity after another has proclaimed that LNG terminals belong in remote places. Jimmy Carter's National Energy Plan of 1977 called for that, as did the U.S. Pipeline Safety Act two years later. California's LNG Terminal Siting Act of 1977 reinforced the concept as well. That act was repealed a decade later; there were no more LNG proposals in the works at that time, anyway, but when applications started reappearing in the last few years, it was evident that many agencies and individuals in California still accepted the idea.

Even the Energy Act of 2005 reiterated the thought very clearly in a list of matters on which FERC should consult with a duly designated state agency. The sixth and final entry in the list is "the need to encourage remote siting."[80] This, one would think, would be clear in meaning, even if not precisely definable. (Dictionaries have no trouble: "located far away," says the American Heritage paperback. My venerable Random House Unabridged gives a couple of relevant definitions: "far apart, far distant in space, situated at some distance away," and "out-of-the-way, secluded.")

It turns out that the concept wasn't mislaid, as at first one might assume by looking at some of the locations where applications for terminals have been accepted, but instead is defined very precisely. If a facility is designed and constructed in accordance with all the regulations, it is by definition remotely sited. The exclusion zones make it so. So says Rob Cupina, deputy director of the Office of Energy Projects, FERC.[81]

The reason for concern, of course, is that nothing is entirely fail-safe. Every EIS, as required by the regulations, addresses the possibility of accident or intentional attack on a facility or a ship. Even the lowest level of consequence suggested has significant impact on the surroundings, and facilities and procedures must be planned accordingly.

That's why there are a number of physical safety aspects to the on-ground design and construction of LNG facilities. First, any LNG storage must have a spill containment system that can contain the entire contents of storage tanks within the area. This may be in the form of a full concrete containment wall around any inner metal tank, with or without a concrete roof, or the tank may sit inside a pit or berm.

Second, there are impoundment areas—ditches and basins in the ground to catch particular, defined spills at potential spillage locations. They are sized to contain the amount of LNG that could be spilled from the largest line within the area flowing full bore for ten minutes (or a lesser duration if the designer can prove a ten-minute spill to be impossible). Always the biggest impoundment area at an import terminal is for the ship unloading line that runs from the jetty to the LNG storage tank, but there may also be another to handle a failure in a pipeline leading to the vaporizers.

Last, there are the thermal exclusion zones, designed to protect the public from pool fires in any of the impoundment areas. These zones are to be controlled entirely by the operator or a governmental agency, and their boundaries are determined by mathematical modeling, based on the area of the impoundment zone. (It doesn't matter whether LNG is two feet or two inches deep within a pool, the fire will be the same size, because only the exposed surface can mix with air and become flammable. The duration of the fire, not its intensity, is what will change according to the volume of the spill.)

The amount of heat that flows through a surface in a specified unit of time is called the thermal flux, expressed as Btus per hour per square foot (or some other measurement of area). LNG exclusion zones are delimited by their potential thermal flux, with the boundary defined at $Btu/hr/ft^2$—this figure, however, is called "low" in the Sandia report. Whether the recipient might consider a second-degree burn to be a "low" consequence is questionable; humans experience second-degree burns from thirty seconds of exposure to a thermal flux of 1,600 $Btu/hr/ft^2$, and that same heat level may cause death in less than a minute. How appropriate that limit is for the exclusion zone has been questioned. "The point of contention," says MIT's Dr. James Fay, longtime critic of LNG facility sitings in urban areas, "is at what distance is thermal radiation intensity low enough that no one will be harmed? What is a safe level to expose people to?" The regulators, says Dr. Fay, are stuck in a time warp. "They use an industrial standard, just like what levels of noise are acceptable within a factory, compared to what's acceptable in your own backyard. This is okay in a factory, but has little to do with protecting the general public."

But not all industrial standards are this permissive. The American Petroleum Institute's Recommended Practice 521 lists standards for exposure to flares within plants, suggesting a maximum thermal flux level of 1,500 $Btu/hr/ft^2$ for "areas where emergency actions lasting several minutes may be required by personnel with appropriate clothing." For locations where personnel are continuously exposed, a level of 500 $Btu/hr/ft^2$ is suggested. The Department of Housing and Urban Development uses the limit of 450 $Btu/hr/ft^2$ for locating outdoor areas—playgrounds, parks, school grounds, and so forth—relative to potential fire locations. The European LNG rule allows a thermal flux at the LNG property boundary equivalent to 1,600 $Btu/hr/ft^2$ in urban places, but only for areas that are easily evacuated on short notice. Other areas are limited to about 480 $Btu/hr/ft^2$.[82]

Many engineers today believe that a flux level in that 450 to 500 Btu/hr/ft^2 range is more appropriate for LNG exclusion zones. For a 74,000-ton LNG storage tank, for instance, FERC's definitions would permit setting the exclusion zone at something like 900 feet. Using a 480 Btu/hr/ft^2 standard would increase the radius of the zone to 1,490 feet.[83] At that thermal flux value, no one would suffer serious injury even if exposed for some time.

Bob Arvedlund, however, defends the 1,600 Btu/hr/ft^2 figure. He says simply, "You'd walk in a different direction. Your skin ain't burnin' or anything like that. And that 1,600 assumes you're nude. You have a layer that's helping you already." To date, the National Fire Protection Association agrees with FERC's former leader, despite the argument that even though most likely they are dressed, not all people have the luxury of moving away easily. The NFPA believes that the present standard strikes "an appropriate balance between safety and the public need for LNG facilities."

And finally, there is the vapor cloud exclusion zone, that area in which, theoretically, the public could be exposed to the potentially flammable vapor cloud developing from an LNG spill. The regulations are not consistent here. Prior to 2000, when the two sets of regulations governing LNG import terminals were melded together, the vapor cloud exclusion zones were calculated for the same spills for which the impoundments were required—in an import terminal, a ten-minute spill from the ship unloading line is normally the largest possible and, therefore, the determining spill. However, due to what appears to be an inadvertent omission of one paragraph during the harmonizing of 49 CFR 193 with NFPA 59A, developers have since that time been allowed to use a smaller "accidental leakage" design spill as the basis for calculating the vapor cloud exclusion zone.[84] This certainly makes it easier to define a vapor cloud exclusion zone that stays on the project's own property, but it defeats the intention of protecting the public. If the possibility of a particular spill has been acknowledged, as it has been by the requirement for impoundment areas, surely then there's an obligation to consider what may happen if that spill should actually occur.

Then comes the issue of the actual mathematical modeling that FERC has allowed for determining the reach of a potentially flammable LNG vapor cloud. (Asked how these models are developed, Dr. Jerry Havens, who led a team that devised one vapor cloud dispersion model and updated another, said simply, "You spend twenty years building a large wind tunnel, and about $8 million of taxpayers' money.")

The first model that Dr. Havens was involved with is called DEGADIS. It was developed in the 1980s, and for essentially flat areas (land or water) it works well to calculate the distances that a vapor cloud would travel while remaining within a defined ratio of gas to air. It does not work in areas with uneven terrain or obstructions such as buildings, dikes, fences, or berms designed to contain spilled liquid.

It had been known since DEGADIS was developed that it wasn't appropriate for use in many of the installations that were being reviewed (during this quiet time for the importation of LNG, these were just peak-shavers and satellite facilities). Dr. Havens and others were called in by the Department of Transportation to research the effects of barriers on vapor clouds. They set themselves to work on what are known as the FALCON tests. They were working under a fellow named Walter Dennis at DOT, who, knowing that the industry couldn't be expected to sit around in the meantime, wrote a letter provisionally allowing the use of DEGADIS in

In the FALCON tests in Nevada in 1987, the experimenters demonstrated clearly that air was mixing with the LNG vapor as it expanded over the fence. COURTESY OF DR. JERRY HAVENS

combination with another model, SOURCE5, to calculate vapor cloud exclusion zones, at least until the results of the research were known.

The FALCON tests determined, among other things, that a key assumption for SOURCE5 is wrong. Gas does not fill a diked-in area without mixing with air. But, says Dr. Havens, "Nobody ever paid any attention to what we found out." Walter Dennis had died in the meantime, but his letter of provisional permission lived on, and still does, nearly twenty years later. In 2000, DOT approved the new model that Havens and the others developed, called FEM3A. It is applicable to spills into diked or impounded areas and deals with other complexities that DEGADIS couldn't handle, but facility developers seem to still prefer the incorrect SOURCE5/DEGADIS method. Perhaps they expect the FEM3A model to come up with longer exclusion distances, but FERC seems to believe the DEGADIS system is actually more conservative.

"Our position is that if it's done according to our requirements, a facility is safe," says FERC's Rob Cupina.

On the water, the issues of concern are a little different. For what it describes as "credible intentional breach and spill scenarios," which involve the release and ignition of half the contents of a single cargo tank, 12,500 cubic meters, the Sandia report says that the distance within which major structural damage to buildings or other vessels could be expected is about 500 meters or 1,640 feet. The distance at which "low thermal hazards" would be experienced is 1,600 meters, nearly a mile. (That is the familiar 1,600 Btu/hr/ft^2 flux level, with second-degree burns occurring in half a minute.)

Presently, no exclusion zones are defined or required around vessels either in transit or berthed. Security procedures nowadays usually call for an area around the ship into which no other waterborne traffic may enter, but other than perhaps some security oversight to protect the ship, there is little if any restriction of activities on nearby land, and therein lies a serious issue about siting. As a ship passes up a bay or river along a populated area, not only might people be at work or peaceably enjoying their own homes within close proximity, but there also might be shoreside facilities that would increase the consequences of an event involving the ship. A petroleum or chemical plant, or even just a lumber dock might add to a fire or to the hazards arising from one.

There's ongoing discussion about which kind of event a plant or ship should be designed for: most likely or worst-case scenarios. According

Artist's rendering of the proposed Cabrillo Port facility shows a tanker tied up to the floating storage and regasification unit permanently anchored offshore. COURTESY OF BHP BILLITON

to the Sandia report, the most likely significant spill that a ship would suffer is a spill from a single tank, but the report recognizes that more than just the single tank might be affected. One can imagine a situation in which a vessel might lose all her tanks. Fire from the first, combined with cryogenic impacts on the structure of the ship, could cause an adjacent tank to fail, and others might follow. Sandia then lengthens its predictions of consequence to 2,000 feet for destruction and 7,000 feet for second-degree burns. So, should the standards aimed at protecting the public be designed around the more probable event or the worst-case situation?

It turns out that Sandia researchers themselves have subsequently envisioned worse cases than they had reported heretofore. In April 2006, the Coast Guard and California's State Lands Commission jointly issued a revised combined draft Environmental Impact Statement for the proposed offshore Cabrillo Port import terminal that may have bearing on other applications elsewhere. Cabrillo Port is a BHP Billiton project, and is to be a floating storage and regasification unit (FSRU) permanently moored twelve nautical miles offshore, two nautical miles from the coastwise shipping lanes. The FSRU is to have a double hull 971 feet long and 213 feet wide. Loaded, the tops of its three Moss spherical tanks would be 161 feet above the water. The single vent stack on Cabrillo Port's FSRU, which would be made up of four- to eight-inch piping and would allow gas to be vented in an emergency or abnormal process, would be the tallest part of the vessel and would stand another hundred feet higher yet. Each of the tanks will store 24 million gallons of LNG, or 90,800 cubic meters, making a total storage capacity larger than the largest LNG tankers on order to date. (In 2006, Qatar ordered a half-dozen "Q-max" tankers with about 265,000 cubic meter capacity apiece to be built at two yards in Korea for delivery in 2008.) This is not a trivial-sized ship. It would be moored far enough offshore, however, that it would be barely, if at all, visible from shore. The vessel would be designed, constructed, and classified under the same standards as a carrier.

Two or three times a week on average, during daylight hours, an LNG carrier would berth alongside the FSRU. She would off-load her cargo around the clock, and the product would be stored in the tanks on the FSRU until it was regasified and sent ashore.

Comments from the first draft EIS and a number of changes in the plan caused a second draft to be produced. The State of California thought the differences significant enough to warrant recirculating the report, although, for some reason, the Coast Guard and MARAD did not. The most significant difference in the revised report, at least as it may relate to other projects, is the new hazards and risk analysis by the Sandia National Laboratories, made specifically for this project.

The volume of LNG on site would be particularly large, though there are other projects considering as much storage. Sandia did say that the consequences from the most treacherous hazard, a pool fire, would not reach the coastwise shipping lane, let alone shore, even if all three of the cargo tanks on the FSRU were to fail. However, they came up with a significantly larger radius of possible impact from a flash fire from a vapor cloud caused by an intentional act than has been suggested in any other EIS. This section of the report has made a big splash among opponents of LNG facilities. While it is very difficult to imagine a scenario in which the tanks could release that much vapor simultaneously without a fire resulting, apparently Sandia felt it could be possible. Unfortunately, their reasoning on this issue is hidden among the security-sensitive information that is kept from the public.

Standards

Both industry and ordinary people who are simply trying to assess the suitability of a particular project for their backyards are frustrated by the degree of secrecy presently being maintained by FERC and the Coast Guard. Of course, one of the main objections to LNG import terminals is the perceived danger to the general public from both plant and attending ships, particularly due to conceivable terrorist activity. Plans must be made to prevent attack, as well as to minimize and offset its impacts should it occur, and, naturally, not all of those plans can be publicized. No one questions that.

Unfortunately, though, more and more aspects of application documentation are now being deemed too sensitive to publish—they're labeled either Critical Energy Infrastructure Information (CEII) or Sensitive Security Information (SSI). CEII and SSI are acronyms everyone hates to see; they are used to justify the omission of whole chunks of information in Environmental Impact Statements. The developer wants the public to feel okay about his project, comfortable that it's not going to cause its neighborhood to go up in smoke, and the neighbors want desperately to believe the same thing. However, much of the information that might allay fears is stashed away from public sight, and anyone allowed access must sign nondisclosure agreements. They are assured that if they blab, prosecution will follow, and they will face a jail sentence and major fines. Most opponents, not wanting to be muzzled, refuse to sign and therefore never get to see the relevant parts of the documents.

One individual who *has* seen some of the secret schemes to handle attacks on LNG carriers assured me that there truly are effective response mechanisms spelled out in the Environmental Impact Statements. Some measures are standard and would be implemented at any facility, others are site-specific. He was personally comfortable with what he knew even though of course he didn't divulge anything even remotely specific about the plans. "They have a high probability of working," he says with apparent confidence. "I couldn't morally support something I didn't believe— you couldn't pay me enough."

There are certainly serious questions as to the safety of proposed facilities, questions that, despite FERC's confidence, some of the most knowledgeable engineers in the field believe are possibly not being addressed sufficiently. In some cases, even if the requirements are adequate, the public has no access to the information, which does not lead to an atmosphere of ease.

Two Contrasting Projects

A S 2007 opened, two of the many LNG projects under consideration were on opposite sides of the country and in vastly different communities. One terminal was proposed for the Port of Long Beach, California, and although as of January 2007 it appears to have been effectively stopped by circumstances unique to that locale, it remains an interesting case study. Some 2,790 miles from Long Beach, as the crow flies, lies Robbinston, Maine, where the formal application process for the other terminal was under way. These two projects represent the extremes in the variety of applications before FERC.

The Downeast LNG project is about as different from Sound Energy Solutions' Long Beach project as can be imagined. Not only are the locations of the suggested facilities antithetical—one being in an industrial city of half a million people and the other in a rural town of five hundred, but so also are the companies themselves. LNG is unique, perhaps, among oil and gas businesses because it lends itself not only to participation by the majors, who own the whole chain from pulling natural gas out of the ground (upstream) to liquefying it, transporting it, regasifying, and ultimately selling it (downstream), but there is also a potential role for enterprising smaller importation operations.

Mitsubishi, the parent company of Sound Energy Solutions (SES), has been dealing with LNG for forty years and supplies Japan with more than 50 percent of its natural gas supplies.[85] ConocoPhillips, which owns a half interest in the California project, describes itself as the third largest integrated energy company in the United States. It has about 35,800 employees in forty countries and assets of $104 billion.[86] By contrast, Downeast LNG was until recently two men—with a group of financial backers. Although he had seventeen years' experience in economic development and the energy field, Dean Girdis was barely forty years old when he founded the company. His offices were for a long time in his attic. He was soon joined by Rob Wyatt, an environmental consultant with particular expertise in permitting, and until the formal application process began, the two of them did almost everything themselves.

Long Beach, California

The effort to build an LNG import terminal in the Port of Long Beach (POLB) provides an interesting picture of both the application process and the issues involved. It became a test case of jurisdiction as well as significant issues of requirements and standards. The facility would have had a capacity of 700 million cubic feet per day and a peak capacity of a billion cf/d. Plans called for the regasification of most of the imported fuel, but some would have been stored on-site and a small amount maintained in liquid form to sell as vehicular fuel. The city has fifty-two vehicles, street sweepers and refuse handlers, that run on LNG.

SES said that the facility could provide about 10 percent of the natural gas needed in California. One of the most significant facts about the Long Beach project, one that somehow didn't always get mentioned by its developers (and appears to have been eliminated at some point), was the plan to take "hot" gas—LNG with a higher than usual percentage of natural gas liquids, propane, ethane, and butane—and remove those more dangerous components on-site for separate sale.

Long Beach, were it anywhere but where it is, would be a big city in its own right, but lying as it does in the shadow of Los Angeles, not many people give it much attention, despite its population of half a million. "Long Beach used to be known as a city where the average person could make above-average income," says resident and anti–Long Beach LNG activist Bry Myown, but when the navy base left, seventy thousand jobs went, too. Immigrants, many of whom live very near the port, have replaced much of the original population.

The port itself is entirely man-made, dredged and filled with "artificial land," as Bry describes it. The area had been just one big, flat flood plain, with a periodic river with no natural bed. "In a wet year, it was like putting your finger over the end of a garden hose," she says. The water ran everywhere. In a dry year, there was no river. Two hundred years ago, incoming ships had to anchor far from shore and lighter their cargo into port. In 1899, construction of the breakwater began, and since that time, dredging and filling have created a port of more than 7,600 acres, teeming with shipping channels, wharves, cargo terminals, rail yards, and roadways. In 1940, the navy opened a base on Terminal Island. Over the years, it expanded from the original 100 acres to 540 before being closed in 1997. SES's proposed facility used a small part of the old naval base.

An unusual aspect of the Port of Long Beach is that it is the site of continuing oil production. Oil was discovered there in 1936. Ever since 1938, pumps have been pulling petroleum from under the harbor waters, and for a time, the oil field was the largest in the nation. After twenty years, a disturbing side effect was noticed: land subsidence. In 1957, a sixteen-square-mile area sank, part of it as much as twenty-four feet. Bry describes the scene: "Plate glass windows imploded, elevators didn't run, and there were surface faults like when mud dries out, only on a much larger scale."

In 1960, the "Big Squirt" project began, pumping water into the subsiding soil to replace the oil being removed. When the oil extraction stops, the water pumping is to stop, too. Says Bry, "Nobody knows what will happen then. No area so large and urbanized has ever stopped injecting water."

Now one of the world's busiest ports, Long Beach receives cargoes from all over the world. Together, POLB and the adjacent Port of Los Angeles handle more than 40 percent of the nation's container cargo (in 2005, more than 14 million twenty-foot container equivalent units' worth, which approaches forty thousand a day).

More than 92 billion dollars' worth of cargo traveled through POLB alone in 2004, some 73.6 million metric tons. Everything imaginable crosses the docks, literally from fruit to nuts, which together appear in the port's top dozen exports by tonnage and by value, along with machinery, plastics, organic chemicals, petroleum products, hides, and various other commodities. Imports include all you'd expect from the East. Though automobiles don't represent the highest value—machinery in general has that honor—they're close to the top of the list. Petroleum products are in the top ten by value and lead the list by tonnage. Eighty percent of the petroleum imports intended for the California market land in Long Beach or Los Angeles.[88]

POLB's Web site claims it generates thirty thousand jobs in Long Beach, which is one in eight local jobs, and 1,400,000 trade-related jobs countrywide. All land, shipping terminals, and other facilities within the port actually belong to POLB, which leases the various facilities to their operators, reinvesting the proceeds in the port.

The LNG terminal was just one of several projects being proposed at Long Beach in recent years; there are also plans for redevelopment of several existing container terminals and, in the next ten or fifteen years, construction of a number of new sites, including four 300-acre container facilities.[87] Each of these projects needs a permit from POLB; the LNG terminal also would have necessitated a modification of the POLB Master Plan, which falls under the NEPA requirement for an environmental impact statement.

In the case of Long Beach, FERC and the POLB agreed to join forces to create a single EIS for the LNG terminal. It's awkward and seemingly redundant, but the jointly prepared document in this case was known as the Draft Environmental Impact Statement/Draft Environmental Impact Report, the first half being FERC's required document and the second being POLB's. We'll call it the DEIS/DEIR, which is awkward enough.

At the time that the LNG terminal was conceived, and even when the application was first filed with FERC, California's Environmental Quality Act required an applicant to have filed for a permit from the California Public Utilities Commission (CPUC) prior to filing with FERC. FERC had proclaimed exclusive jurisdiction over LNG terminal siting, however, and in response, the CPUC and the Californians for Renewable Energy (CARE) filed suit in the Ninth Circuit of the U.S. Court of Appeals, hoping to clarify whether a state had the right to effectively argue against the suitability of the project's proposed site location. Could the CPUC stop a siting approval issued by FERC?

In August 2005, in the midst of the court proceedings, Congress passed the Federal Energy Policy Act of 2005, EPA05, proclaiming FERC to be the lead agency on all shore-based applications across the nation and giving FERC the sole authority to approve a project's site location. At that time, the CPUC threw in the towel, feeling their position was moot, but CARE kept on. Although ultimately their first filing to the U.S. Circuit Court was kicked out, CARE did seriously consider filing an appeal with the Supreme Court, though they were not terribly hopeful that their case would be heard.

Many people in Long Beach and elsewhere in the state at first wanted this project to go through, including the city's mayor, a majority of the city council members, the five-member Board of Harbor Commissioners (appointed by the mayor), and Governor Schwarzenegger. Many others were critical of it from the start. Some believe LNG is too dangerous to mess with, period, and a variety of other objections were also raised.

There were the predictable greens of one stripe or another, many of whom would have been against any LNG imports at all. One of their major arguments is that California doesn't need additional natural gas supplies, and adding more natural gas to the supply now would simply increase dependence on foreign energy sources while at the same time undercutting initiatives for developing sources of renewable energy. Importing more subsidizes inefficiencies, the greens say, and encourages increased gas uses such as producing more plastics (which then must be dealt with as solid waste).

By state law, 20 percent of California's electrical energy generation is to come from renewable sources by 2017, and, in fact, they're hoping to meet this goal by 2010. The state is tied into firm, long-term contracts for both coal and nuclear energy, so the only part of its portfolio that can be replaced by renewable energy is the natural gas sector. Today, about 30 percent of the state's total energy comes from natural gas, and 42 percent of the electricity generation in-state is fired by gas. Actual production of natural gas within California is but a sixth or so of the overall need and is expected to remain flat. A quarter of the gas used comes from Canada and the rest from the Southwest or the Rocky Mountain basin. The state projects a smaller increase in overall demand for natural gas than in the country as a whole: 0.7 percent per year through 2016 compared to 1.7 percent nationally. The difference in projections is particularly striking for power generation: the U.S. Department of Energy expects natural gas usage for electric power generation countrywide to increase 4.3 percent annually through 2016, whereas California's projection is for only a 1 percent increase each year. The state energy policy from 2005 does call for importing LNG, however. [89]

What is interesting, according to Bry Myown of Californians for Renewable Energy, is that in California's natural gas–fired electrical generation plants, more fuel is presently being wasted—literally blown off into the atmosphere (and contributing unnecessarily to global warming)—than would have been imported by SES. She, and environmentalists in general, suggested it would be wiser to upgrade the existing plants.

Other objections to the import terminal included a number of concerns about safety in this particular location. For decades, California has taken the position that LNG terminals should be placed outside of areas of high population density. As discussed, the Energy Policy Act of 2005 still refers to remote siting. The definition of that term has yet to be established in court, and only a definition like FERC's would allow POLB to have a "remote" site within it.

The proposed plant was to be built on only twenty-four acres, the smallest site for any LNG terminal in the United States. Immediately adjacent to the proposed site is a pier used to unload lumber and oil, both flammable materials. Within a mile is the Defense Logistics Agency Fuel Facility, which provides jet fuel to the military in the Southwest as well as to navy ships. Maritime Prepositioning ships are stationed in the vicinity, and the harbor fire department headquarters is also nearby. Inside a one-mile radius are two thousand port workers.

The remainder of the port holds many more facilities with their own intrinsic hazards: chemical plants, oil tanks, pipelines, and refineries. There are also tourist attractions such as the *Queen Mary*, which several thousand tourists may visit on a summery day, and terminals for Carnival Cruises and the Catalina Express boats that run out to Catalina Island.

In addition, within three miles of the site, between the port and the downtown area, 44,000 people work and 85,000 make their homes. An interesting wrinkle is that the surrounding population is disproportionately made up of members of minority groups and impoverished people. According to the Census Bureau's American Community Survey, the percentage of the Long Beach population with incomes below the poverty level from thirty-seventh in the nation in 2000 to sixth in 2005. Near the harbor live people whose median household income is about a third less than that of people living just a couple of miles farther away. Sixty-five percent of the residents of this area are of Hispanic origin. One might wonder why those socioeconomic facts would matter, but one of the objections to the facility was that it would be in violation of state and federal civil rights guidelines because it would have a disproportionate impact on the minority community.[90] The aftermath of Hurricane Katrina certainly demonstrated how disaster hits impoverished people the hardest; many argued that the potential for disaster from the LNG terminal's location in POLB constituted a socioeconomic injustice.

There were geological problems to be considered, too. The site is near a number of seismic faults, one of which produced a magnitude 6.4 earthquake in 1933. SES's own Resource Report No. 6 identifies twenty-seven faults "potentially having a significant contribution to the ground-motion hazard at the LNG terminal site." It is also in a zone where soils are subject to liquefaction, which greatly increases the likelihood of building damage or collapse during an earthquake. Elevation of the port area is at or even below sea level, which means the site is in a flood zone and, due to the potential for earthquakes, is particularly susceptible to tsunamis and seismically induced flooding.

Having been designated by Governor Schwarzenegger as the lead state agency to consult with FERC on this application, as the Energy Policy Act of 2005 allows, the California Energy Commission reported its concerns to FERC prior to the publication of the DEIS/ DEIR. Among many other issues they raised was that of the seismic conditions. But for some reason, the physical issues didn't seem important to the third-party consultants who wrote the DEIS/DEIR. (A major point the California Energy Commission made in their official comment to the draft was that FERC still needed to address the issues they'd raised earlier.)

SES's application for Long Beach originally offered a scenario different from that of many LNG import facilities, because they intended to import "hot" LNG—product containing a higher than normal proportion of the heavier natural gas liquids, propane, ethane, and the like. They later filed a number of revisions to their application dealing with these other components, but apparently always planned to be able to handle them in some fashion.

Despite the expressed concerns of a growing number of people from a widening spectrum, the application process continued. Published in October 2005, FERC's Draft Environmental Impact Statement and Report ran more than 350 pages, not including an eleven-page table of contents and a five-page supplement (single-spaced) defining all the acronyms and abbreviations that appeared in the report. (It was, after all, an opus for the government.) That tome did *not* include a number of sections still to come at some unspecified time in the future. Portions designated as Critical Energy Infrastructure Information or Sensitive Security Information were not disclosed, of course, and certain sections regarding public health and safety also were deemed sensitive information and not made available for public view, which raised some furor as well.

As is always the case, the overall conclusion of the draft statement was favorable toward the project. The comment period lasted until December 8, and many individuals and organizations filed comments, some supportive, some critical.

Predictably, environmentalists continued to question the need for importing additional natural gas in the first place. The interesting thing about the Long Beach saga was that by the time the DEIS/DEIR was issued, the green groups had some important company supporting their objections to the project. Among many other organizations and individuals taking serious issue with the report was the California Public Utilities Commission (CPUC). In addition, the California Coastal Commission and the California Energy Commission and various other establishment-type groups all objected to at least parts of the DEIS/DEIR and filed strongly negative comments with FERC.

The CPUC pulled no punches, calling the Long Beach location "one of the worst possible sites imaginable." The Coastal Commission, which has the task of protecting both environmental and human-based resources along the California coast and ocean, was hardly any gentler, stating that the draft document was "fundamentally inadequate" in a number of aspects, starting with its failure to identify and address a number of safety issues. The California Energy Commission (CEC) filed a line-by-line listing of the deficiencies and inaccuracies of the report. Even some organizations that did not take a position one way or the other in their comments, such as the City of Long Beach itself (previously in favor of the project), stated that they believed the draft document was inadequate because it did not consider various significant issues. The city councilors appeared to have become less enthusiastic about the project, and members of the staff also made their displeasure known.

Inconsistencies and discrepancies within the report itself were pointed out, both simple ones, such as the number of trucks entering the site during peak construction hours (twenty-seven per hour, or eighty?), and critical figures such as those regarding the impact of temperature at various distances during a spill and subsequent fire. These calculations *are* subject to argument, and plenty of argument has been offered, but it is reasonable to expect consistency within the document.

Among the serious omissions and errors in the draft report that commenters focused on were the lack of any information or analysis of the consequences of a release from the onshore natural gas liquids storage tanks or pipelines, the lack of consideration of navigation issues within the harbor, the incorrect baseline for comparison of one project to another, the lack of attention paid to seismographic conditions, an inadequate analysis of the air quality impacts of ships and tugboats, and the misuse of one of the mathematical models used to determine the exclusion zones around the facility. Also, several filers of comment stated that the report did not meet the requirements of either the National

Environmental Policy Act or the California equivalent, the California Environmental Quality Act (CEQA).

Among the issues raised from a number of quarters were questions about assumptions made regarding the appropriate location of an import facility. For example, the authors of the CPUC's major response to the DEIS/ DEIR began by recognizing the need for LNG import terminals, then they noted the hazardous nature of LNG and stated that the potential consequences of a spill could be very widespread. "Therefore," they concluded, "where, as here, there is a choice between an LNG import terminal in a densely populated area or [at] a remote onshore or offshore location, common sense, and the law, would require that the LNG import terminal be sited in the more remote location."

The CPUC also objected to the DEIS/DEIR's failure to address a needed pipeline expansion (which the CPUC said the developer deliberately concealed) or the issues surrounding the storage and transportation of the propane extracted from the imported LNG. The CPUC authors also questioned a number of claims made by SES about potential impacts from earthquake, terrorism, fire, or explosion, and about capacity of the existing pipeline system. The CPUC said these called SES's credibility into question. The DEIS/DEIR had been issued prematurely, concluded the CPUC comment, as it neither answered questions raised in the state Safety Advisory Report, as it was required by law to do, nor complied with the requirements of NEPA or CEQA. The required Port Master Plan Amendment would also violate California law in a number of ways, mostly relating to the proposed terminal's urban location, concluded the CPUC, and should be turned down.

Almost funny were the DEIS/DEIR calculations of the probability of a catastrophic event, which nearly everyone agrees would be most likely to come at the hands of terrorists rather than by accident. "It is impossible to predict the probability of occurrence of specific intentional events (such as those perpetrated by vandals or terrorists)," wrote the experts from Quest Consultants, Inc. But they immediately went on to calculate that probability to be seven in a million per year. They supported this calculation by analyzing figures from an EPA database of facilities with threshold amounts of toxic or flammable materials. The EPA says there are 12,711 such facilities where a catastrophic event could affect 100 or more people (itself an arbitrary criterion of questionable pertinence).

Quest did its study in 2004, eleven years after the first high-profile terrorist event on U.S. soil: the February 1993 bombing of the World Trade

Center. No successful terrorist act actually occurred in any of the 12,711 defined facilities during that eleven-year interval, but in order to be able to calculate the odds of such a thing occurring, Quest started with the assumption that one successful terrorist event *had* taken place. One event in 12,711 facilities in eleven years equals one event in 139,821 facility years. This led them to conclude that there is about a seven in one million chance of a terrorist-induced failure occurring in any one of the 12,711 facilities in any one year. (By some accepted fiat about risk assessment, this logic apparently is deemed acceptable).

Interesting to the CPUC, the California Energy Commission, and many others was SES's analysis of alternatives. In this DEIS/DEIR, alternatives to the project at hand were compared not for their relative contribution to the greater good, but instead, primarily in terms of how well they would meet the applicant's expressed goals. Since Sound Energy Solutions' goals included that SES itself bring natural gas to the southern California market, any alternative that didn't allow it to do just that would, by definition, be substandard. "'No Action' or 'No Project Alternative' would eliminate the environmental impacts identified in this EIS/EIR," observed the DEIS authors, but then "none of the objectives of the proposed project would be met. Specifically, SES would not be able to provide a new and stable supply of natural gas and LNG vehicle fuel to southern California." To be fair, they went on to acknowledge that there were other proposals that might be able to provide a new source of natural gas to southern California markets. "However," they said, "none of these system alternatives could meet the proposed project's stated objective of providing a stable source of LNG for vehicle fuel."

This was simply untrue, as filed comments noted. Other proposed projects could provide LNG for vehicle fuel, as could existing liquefaction plants in the state, if doing that were, in fact, important. Its importance even to SES would have been minor, the annually promised amount of vehicle fuel being equal to only 1 percent of the total import from a single ship. But the fact is, as nonprofit FuturePorts pointed out in their filed comment, there is little if any advantage in using LNG to fuel vehicles, anyway. The desired reduction in particulate emissions can be achieved more simply by utilizing trucks built or upgraded to fully benefit from the new ultra-low-sulfur diesel fuel. If the goal were simply to decrease emissions, which California is adamant must be done, SES's proposal for LNG fuel didn't even do that.

As for their competition, the DEIS/DEIR dismissed offshore LNG facility alternatives with various arguments, which were debunked one by one by the offshore applicants. In its filed comments, Crystal Energy, developer of one of the offshore projects, said that weather causes more problems for boarding pilots and Coast Guard officials than anything else, and those difficulties apply to any vessel, whether she's headed to an offshore or an onshore unloading facility. Other arguments about the impact of weather, said Crystal, should not be speculated on, as the DEIS/DEIR lacked factual support for its statements. And even the final DEIS suggestion that SES's application was superior to the offshore facilities for being in an already developed industrial area could be balanced, at least in Crystal's case, by the fact that Crystal was proposing to build on an existing but now inoperative oil platform in the Santa Barbara Channel.

The California Public Utilities Commission, like many other commentators, gave the largest amount of space to issues of safety and security. They looked to Dr. Jerry Havens, long considered one of the foremost experts in the LNG field, to articulate one of their most serious concerns: what about the consequences of a terrorist act that could cause spills of three million gallons (or even considerably more) of LNG onto the ocean? The Sandia report saw such an event as credible, but this DEIS/DEIR did not even consider the possibility.

"A *minimum* 3-mile radius circle around the proposed terminal is proposed to demarcate the area in which events deemed credible could cause injury to the public," Havens suggested, adding that such events would be likely to cause cascading failures both on the vessel and within the port, and that the consequences could be even more severe than he'd described.

The CPUC also made note of what might arguably be the most significant aspect of the proposed Long Beach siting: "The DEIS/DEIR never addressed the devastating economic impact that could occur [from an LNG event] in light of all the commercial activities and critical infrastructure already existing in the Port of Long Beach."

The end conclusion of the DEIS/DEIR was simple. "SES's proposed project is the environmentally preferable/superior alternative that can meet the project objectives." Of course, there were several pages of requirements SES would have to meet before and during construction, and forever after during operation of the facility, but the report would lead one to believe that downtown Long Beach is not only a fine place to put such a facility, but also the very best place.

The conclusion of the CPUC comment was equally straightforward. They considered the potential peril from human error, earthquake, or terrorist event as extending at least three miles out. "When human error alone makes this risk too large in light of how many people would be in harm's way (i.e., at least 130,000), the added risks of earthquakes or terrorist attacks makes this site one of the worst possible sites imaginable."

Debate over the proposed Long Beach facility continued, seemingly unendingly. It looked as if the issue would ultimately be decided in court (and that may still happen). Annoyingly to the objectors, deadlines for various aspects of the application came and went without being met, and no one seemed to mind. A revised draft was not circulated, as so many agencies and individuals felt would have been appropriate, and the final EIS was delayed several times. Despite the system for filing comments with FERC, many believed that none of the comments actually received attention from the commission. "We don't have any venue where we get heard at all," said CARE's Bry Myown. When the movie crowd in Malibu began protesting the offshore projects near them, some Long Beach LNG opponents worried that FERC would see that as evidence that no one wants LNG anywhere, anytime, and would therefore overrule *all* objections—"so we don't freeze in the dark," as one interested person put it.

Interestingly, it is a local board, not the federal regulators or state commissions or activist citizen groups, that appears to have stopped the Long Beach LNG project in its tracks. The city of Long Beach is unique in that its harbor commission serves as landlord for all harbor projects. The harbor commissioners, during the years of debate and discussion, gradually became convinced that an adequate Environmental Impact Report was not forthcoming, and in late January 2007 they voted to disapprove the project. As this book goes to press, the Long Beach LNG terminal appears to be dead, although there is still a possibility that SES may take the city to court.

Robbinston, Maine

On the easternmost edge of the United States, the little Maine town of Robbinston sits on the edge of Passamaquoddy Bay overlooking St. Andrews, New Brunswick, Canada. With a population of five hundred, Robbinston is a town where not much goes on; indeed, all of surrounding Washington County ("the Sunrise County"), is equally quiet. Some people like it that way, while others wish their sons and daughters didn't have to leave home to make a good living. Not everyone wants to go lobstering, and only a certain number of teachers, real estate brokers, waitresses, and woodcutters are needed in any small community. Census figures demonstrate that residents of Robbinston and Washington County are considerably older, less educated, and less well-off than those in the state as a whole.

Because of the need for natural gas in southern New England, the entire coast has been scrutinized for possible LNG import terminal sites. ConocoPhillips was the first company to announce a project in Maine; they intended to build on property they would lease from the southern Maine town of Harpswell, in Casco Bay. The company is accustomed to doing business in a large way, and has been accused of coming into Harpswell with a preconceived plan that they expected to simply set in place and call good. However, a majority of Harpswell residents didn't see it the same way, believing that LNG was probably dangerous and certainly contrary to their way of life on their narrow peninsula in the busiest bay in Maine.

Casco Bay is the site of Portland, the third largest petroleum port on the East Coast and a city already accustomed to seeing both cargo and passenger traffic in significant quantities. Casco Bay is at some seasons speckled nearly solid with neon-painted lobster buoys, each attached to a gang of eight to ten traps, and it is also the playground for the myriad pleasure boats of summer visitors. Seasonal homes line its shores. ConocoPhillips's Harpswell project was in fairly short order trampled out by the local community. (Not without a lot of rancor and heavy disruption, even within friendships and families, the shadows of which continue to this day.)

Various companies expressed interest in other locations in Maine and other New England areas, and, in fact, applications were made for a number of sites, onshore and offshore. One of these, the KeySpan plan for Providence, Rhode Island, has the honor of having been the first (and so far only) application to be turned down by FERC. Interestingly, in the press release following an appeal of their decision in that case, FERC acknowledged two reasons for the turndown: inadequacies regarding the existing storage tank, and (perhaps important in terms of setting a precedent) the fact that "the thermal radiation and flammable vapor exclusion zones would extend offsite onto adjacent properties."[91]

Three separate proposals for far down east Maine have been discussed. Passamaquoddy Bay offers deep, protected water in which ships can unload comfortably, it is reasonably close to the Maritimes & Northeast pipeline that already carries natural gas across Maine to the Boston market, and it is in an area in which much concern has been expressed about the need for economic development. The low population density there can be seen as an attractive characteristic from the point of view of security. What terrorist would want to cause trouble in an area where with the greatest of luck he might be able to kill twenty members of the general public and some seagulls?

The first-mentioned Passamaquoddy Bay plan came from Quoddy Bay LLC and was shot down by vote of the townspeople of Perry, who, it turned out, had approval authority over commercial development of the particular parcel of land under consideration.

Quoddy Bay LLC came back with another plan, this time on land owned by the Passamaquoddy Tribe on Pleasant Point, near the town of Eastport. They won the race, if race there was, to get the first letter to FERC requesting pre-application for a new LNG facility in Maine, though Downeast LNG, with its Robbinston site, was only a couple of weeks behind them. The third group, attempting to develop a plan for a site in Calais, has, at this writing, yet to make any filing.

Dean Girdis is Downeast LNG. His background is in economic development and energy, starting with a Peace Corps tour in Mali and continuing with seventeen years in international consulting. Starting in 1997, he was asked by the World Bank to analyze the potential for LNG in China. Would it be a competitive energy source for power generation? What planning needed to be done? What were the environmental issues? What were the pertinent legal issues? China is completely different from down east Maine, or Long Beach; the local communities have absolutely no say about anything to do with industrial development. If China's central government wants an LNG plant, that's that. In the course of working on this Chinese project, Dean met the players from the big oil and gas companies: ExxonMobil, BP, Shell, and other private-sector LNG people. Some of them thought he was from Mars, he says. His background was with the Peace Corps, he had no car, and he ran fifty-five miles a week. But he knew what he was talking about, and he became interested in regasification.

Watching ConocoPhillips in Harpswell, he could see they weren't going to make it work. Originally from New England, he felt he could understand the local issues better. "If you're from Texas, you don't get the same reception," he says. Looking to Cheniere's experience on the Gulf Coast as a model, he thought he could do an LNG project in Maine himself. In March 2004, he founded his company, Downeast LNG; its office was in his attic in Washington, D.C.

It wasn't long before he asked Rob Wyatt to join him. A professional environmental consultant for thirty years, Rob had expertise in the permitting process that is such a huge part of any energy project. (He has the distinction of having been the first person to get an American LNG import facility permitted since the 1970s, the EcoEléctrica plant in Puerto Rico.)

By networking, Dean found investors. Kestrel Energy Partners signed on with him, understanding that while certainly a respectable profit might be made, there was nonetheless a 70 to 80 percent chance of losing their entire investment. Downeast LNG was real, with both large ideas and financial support.

One of the first documents Girdis and Wyatt produced was a Corporate Responsibility Statement, announcing that Downeast LNG was committed to responsible conduct, responsibility to the community, and responsibility to the environment. Anyone who spends any time with Dean Girdis knows that he's going to be open; he says what's on his mind. He may moderate his vocabulary some when he's speaking in public, but the words fall out as fast as words can, his tongue attempting to keep up with his brain. Wyatt has spent time in the South, one can tell, but he's quick to say what's on his mind, too, and prides himself on being forthright.

After listening to the concerns of the people of Robbinston, they worked up a contract by which Downeast LNG will provide the town not only with considerable property taxes and the additional infrastructure it needs should an LNG terminal come to town—added fire protection and other emergency equipment, for instance—but also with a new K–8 school, a policy of priority in hiring, compensation for demonstrated losses to businesses or fishermen due to the company's operations, homeowners' compensation for near neighbors, and an ongoing contribution of more than a million dollars a year to a community development fund to be used as the town sees fit.

Toward the end of 2005, two petitions were handed in to the selectmen of Robbinston, the local official board responsible for running the town. One, bearing 190 signatures, asked that a vote be held as soon as it could be arranged in order to demonstrate the townspeople's support for the project, or lack thereof. The second, with seventy signatures, asked that any vote be postponed until after more information was in hand. The selectmen opted to go with the request that more townspeople chose to sign, and scheduled a nonbinding vote for January 10, 2006. A public hearing on the question, held during the last days of 2005, attracted about seventy of the five hundred residents of Robbinston.

So they could report back at the public hearing, a group of townspeople traveled to Maryland to check out the Cove Point LNG facility beforehand. The plant manager gave them a tour and told them that the plant had recently had its first loss-of-work incident in six years—from a bicycle accident. (The Cove Point ship's berth is a mile from shore, connected by a tunnel that carries both piping and bicyclists out to the dock. A certificate of accomplishment is given to people from outside who pedal all the way through the tunnel.) "We saw a fail-safe system," reported Mike Footer of the visiting group, though he admitted that, of course, it could fail. "The space shuttle failed. But," he said, "there are a lot of interlocks—if this happens, then that'll happen." He compared the plant to the Domtar paper plant twenty-three miles up the road, at Baileyville, at which he and a number of other Robbinston people worked. "The computer does shut the place down sometimes, just like at Domtar, and it'll come back up again and you go on your way."

Cove Point was a clean, quiet plant, and its impact on the Maryland community appeared to be minimal. In fact, when the group asked people they met on the street how they felt about the LNG plant, most of them didn't even know it was there.

At the Robbinston hearing, a heavyset man with his dark hair cut in a flattop stood up. "I have two questions," he said. "Were you attacked by terrorists while you were there, and if not, how many battalions of marines protected you?" Everyone laughed, and then the research group reported that they encountered more rigorous security when entering the LNG plant than when going through the airports on their way to Maryland. "We had to send in our names and social security numbers ahead of time, and we went through a metal detector. 'Course, we stood out," said Mike. "I don't think they get many visitors there."

Although it was clear that they were outnumbered in the room, a few Robbinston residents spoke against the project at the hearing. The threat to what they saw as the pristine nature of the bay was a concern for some, but they were primarily concerned about their own property values and insurance rates. Some questioned the timing of the vote and whether it would be the last opportunity for townspeople to have a say. (They were assured it was not.) Ten days later, the polls at the fire station saw more voters than had turned out for any previous election. Two hundred twenty-seven of them supported the project and eighty-three voted against it. One hurdle was past, and it appeared Dean was right: if you work with the people of an area, you can accomplish a lot. He describes his project as a partnership with the people of the town, and they appear to have accepted that.

In late December 2006, both Quoddy Bay and Downeast LNG filed their formal applications with FERC. By that time, Downeast had generated 300,000 pieces of paper documenting and supporting their project.

It is possible that the most serious block in their way might come from the Canadian government, which may attempt to stop LNG traffic from traveling through the narrow Head Harbour Passage into Passamaquoddy Bay. The messages received to date on the subject are distinctly mixed. The Canadian technical agencies are said to be working in cooperation with the U.S. Coast Guard to define waterway use rules for the LNG ship transits, and were involved in a joint two-day conference significant to the Coast Guard approval process for the Downeast project. (Similar cooperation has occurred in other border areas, including the St. Clair River at Detroit and the Strait of Juan de Fuca in Washington State, where Canadian vessels pass through American waters on their way to Canadian ports.)

Still, there is serious local opposition to the Robbinston project because of the concern about the environmental impact to the passageway, and the residents of St. Andrews feel it will have a negative effect on their tourism.

The final EIS is supposed to be out in the summer of 2007, and perhaps the certificate (Approval of Place of Import for Natural Gas) would be in hand by fall. Construction could conceivably start by the end of 2007. Many such approvals are challenged in court, however, and it could be considerably longer before construction could begin.

Different Regions, Different Results

THAT New England could make good use of more natural gas than is presently available is undisputed. The nation's cold northeastern corner has neither California's economic clout nor its sun. The small New England states can't demand anything from anyone, the way California mandated lower automobile emission levels from the manufacturers. They can't provide market enough to make it worth anyone's while to develop serious alternative electrical energy generation sources, either. There's no coal here, no oil. Nuclear power plants are in use but no new ones are likely to be built in the immediate future. Even in the best of situations, solar energy has limited potential in this region. Imported LNG has clear appeal for New England, and the Maine and the offshore Massachusetts applications already mentioned are not the only projects to have been proposed to date.

Two more LNG facilities were proposed close to one another in southern New England, one in Fall River and the other Providence, Rhode Island. Both were relatively early coming into the FERC review process. In June 2005, one was approved and the other was turned down. Both decisions were appealed; both were upheld. Providence may possibly be finally dropped, but in October 2006, a combined petition from all opponents to the Fall River project was filed with the U.S. Court of Appeals.

In the northeast corner of the country, as on the West Coast, the subject of importing LNG seems to remain as controversial as ever. Look to the Gulf of Mexico, however, and you find a different picture. The applications for new LNG facilities in Louisiana and Texas stirred up relatively little public opposition.

Fall River, Massachusetts

According to the last census, Fall River is a city of somewhat more than 91,000 people, about 1 percent fewer than ten years earlier. It is an hour from Boston and somewhat closer to Cape Cod and to Newport and Providence, Rhode Island. For the twenty-odd years before and after 1900, it was the center of the nation's cotton textile manufacture; now, many of its big granite and brick mill buildings are home to offices and other uses. Some still lie vacant.

Fall River stands out in an unfortunate way on maps depicting municipal family poverty, household income, education, housing value, school dropout rate, and even SAT scores.[92] Incentives exist for redevelopment, but the city has a ways to go to catch up with its wealthier neighbors.

Gordon Shearer, who worked at Distrigas under Samuel Bodman (now U.S. Secretary of Energy), heads up Weaver's Cove Energy, which is proposing to build an LNG terminal on the site of an old oil tank farm just north of downtown Fall River. The site has some advantages: it has deepwater access, it's already industrial in nature and is zoned for marine industrial use, and it is close to a natural gas pipeline and to an interstate highway, which would provide access for trucks carrying LNG to peakshavers and elsewhere. The entire region around Fall River will benefit from the new facility, says Weaver's Cove, because not only will they clean up the defunct oil storage tanks and the ground contamination presently on-site, but they will also dredge the river, which would benefit other marine traffic.

There are also disadvantages to the site. Eight thousand people live and work within a mile of it, and nearly a half million people live in the communities abutting the tankers' route. It is twenty-six miles from the open ocean, and the ships would have to pass under four bridges to reach the site. Twenty-three miles of the voyage are through Rhode Island's Narragansett and Mount Hope bays, both crowded waterways with densely populated shores. Narragansett Bay is home to much of the commercial and recreational water traffic of Rhode Island, which counts more than 60,000 registered boats.[93] In addition, Rhode Island waters are crisscrossed by ferry traffic, tour boats, sailing regattas, and many commercial fishermen, and Newport is a stopover for cruise ships. International events like the Tall Ships gatherings take place there. All of these would be inconvenienced, if not driven away, by the frequent passages of ships requiring extensive security zones.

As a consequence, Fall River's mayor, Edward M. Lambert, Jr., and the attorney general of Rhode Island were sufficiently concerned about the possible safety issues that they asked Dr. Jerry Havens to assess the two LNG projects' safety implications. "I concluded that the concerns that had been expressed initially to me were more than well founded," Havens testified to FERC. He questioned the calculations of impoundment sizes and exclusion zones. Theoretically, both are sized to protect the public, but as we've discussed, vapor cloud exclusion zones are calculated from smaller spill quantities than would be held in the containment basins in the event of a serious leak. Dr. Havens explains. "You are required to provide a spill basin for collecting the full line spill from the biggest line going full blast for ten minutes. But you are not required to plan for the vapor cloud that would result from that spill." Such a cloud, were it from a major spill that filled the basins, would extend beyond the Weaver's Cove site. That fact alone should preclude approval of the facility.

An interesting aspect that points to disingenuousness on someone's part is the matter of the models used for calculating vapor cloud exclusion zones. Dr. Havens reported to FERC that the FEM3A model would be more suitable for this application than the DEGADIS/SOURCE5 method actually used in Weaver's Cove's EIS. In a rebuttal to Havens's comments, the lawyers for the developers stated that the FEM3A model was not available. The mayor's office was able to ascertain that not only was the model available (which I confirmed by a simple phone call), but Weaver's Cove's attorneys knew this—they had been told so by the Gas Technology Institute, which distributes the model.[94]

Dr. Havens was also concerned about the lack of exclusion zones around ships, even though there is no disagreement about the potential for serious consequences from accident or terrorist attack on a vessel. This subject was discussed in tedious repetition in fifteen months' worth of filings and comments following a petition by the City of Fall River to the Coast Guard to establish the exclusion zones around ships both in transit and while moored. On the one side were the City of Fall River, the attorneys general from both Massachusetts and Rhode Island, and various now-predictable players such as the Conservation Law Foundation and local nonprofit and lobbying groups. Senators Kennedy and Kerry and the Massachusetts congressional delegation joined them. On the other side was an array of equally predictable parties, including nearly every company big and small within the world of LNG, along with such industry organizations as the Center for LNG and the Society of International Gas Tankers and Terminal Operators (SIGTTO), all of whom said the exclusion zones weren't necessary, that the industry had a proven record of safety and very robust vessels, and it shouldn't be treated differently from other transportation sectors. Mitigation measures to reduce the threats, vulnerabilities, and consequences could be implemented and would suffice to protect the public. Both sides referred to the Sandia report to support their positions.

Ultimately, the Coast Guard turned down Fall River's petition.

During this spate of comment filing, Patrick Lynch, Rhode Island's attorney general, contacted security expert Richard A. Clarke to ask whether terrorism was a real consideration for the LNG project proposed for Providence. Before testifying before FERC, Clarke analyzed the risks of terrorism posed by projects of the type being considered for both Providence and Fall River. He flew over the project areas and their approach from sea. When it came to siting a facility in Fall River, his conclusion was clear. "I can think of few actions that our government could take that would be as prejudicial to the public." In his testimony before FERC, he stated, "An urban LNG facility would rank high on any terrorist's list of target opportunities."

Local opposition to the Fall River terminal continued to grow. There were ten to twelve thousand names on one petition against the project. An organization called Friends of LNG did form in support of the project, saying it represented some five hundred members. The developers claimed no connection with the group, until the local paper discovered that Weaver's Cove LNG supported the group financially. Asked for a list of members, the spokesperson of the group did not produce one, leading some wags to thenceforth refer to the organization as "Friend of LNG."

Concern over the lack of exclusion zones notwithstanding, FERC granted its approval to Weaver's Cove at the end of June 2005, on a vote of three to one. It is unusual for FERC to have a split vote, and, in fact, this was their first split decision ever pertaining to LNG. Following FERC's approval, the City of Fall River, the Rhode Island and Massachusetts attorneys general, and the Massachusetts Energy Facilities Siting Board jointly filed petitions for a rehearing. At that time, the U.S. Navy made its own request asking that the process be reopened.

Originally, the navy had objected to the project because of the proximity of the tanker route to a Naval Undersea Warfare Center where underwater weapons testing takes place. A week before the FERC commissioners' meeting to consider a rehearing, the navy changed its mind and withdrew its objection. It had received assurance from the Coast Guard's captain of the port, Captain Roy Nash, that the navy would not be required to honor the safety and security zones around a transiting

LNG carrier. "However, all parties remain responsible for the safety of their operations," wrote Captain Nash.[95] (One has to wonder what are the implications of this statement.)

Eric Poulin, assistant to Mayor Lambert, believes that concerns expressed by any department under the federal executive branch simply disappear. "They're all marching in a line, and no one knows who's running it," he says. An occurrence in San Diego would seem to support this view. When it was suggested that an LNG terminal be built near Camp Pendleton, the commander of that Marine Corps base came out against it. Within a week's time, however, the head of Marine Corps installations articulated a revised position. (It's not entirely clear whether the marines' position was completely reversed or merely made more tactful.)

An interesting aspect of the Fall River situation is that everyone is working together to defeat the project: Republicans and Democrats, two states, and several municipalities with little history of joining together, particularly not on an issue that they might dismiss as being simply Fall River's problem. Of course, Weaver's Cove isn't just Fall River's problem, or even Massachusetts's, because the ship transits through so many miles of Rhode Island waters. And then there are the bridges.

Bridges are a part of life in Fall River. South of town, the Braga Bridge takes Interstate 195 across the Taunton River estuary where it meets Mount Hope Bay. A mile or so north, the old Brightman Street Bridge has linked Fall River with Swansea since 1908. For many years, the city lobbied to replace the old drawbridge. It stood only twenty-seven feet above mean high water and was having to be opened constantly to let vessels through—more than a thousand times a year. It was costing too much to keep in repair. Finally, in 1998, construction began on a new bridge. By the time Weaver's Cove LNG facility came along—the first paperwork was filed in 2003—the expectation was that the old bridge would be torn down and gone before the project was built; indeed, it should have been destroyed already, but contractor errors had delayed completion of the new span. "But that worked out to our favor," says Eric Poulin. If the old bridge were to remain in place, it would stop the LNG traffic, for the 145-foot-wide ships that would service the proposed facility couldn't fit under the old bridge's 98-foot span.

In August 2005, Massachusetts congressmen Jim McGovern and Barney Frank inserted a clause in the Federal Transportation Bill that proclaimed the old bridge to be an economic development tool, providing a bikeway, pedestrian walkway, and fishing piers that would draw people to the city. The wording of the bill was drafted not only to preclude the use of any federal money to tear down the old drawbridge, but to earmark a half million dollars for improvements to it. The state had already set aside $5 million for the bridge's demolition; now Fall River hopes that funding will be made available to benefit the bridge instead, and a task force has been formed by the Massachusetts Highway Department to discuss these issues.

The opponents of the LNG facility, including the city, were thrilled; they had stopped the LNG project. Weaver's Cove offered to pay for demolition of the bridge, but that was not an offer anyone wanted to accept.

The request for a rehearing came before FERC in January 2006. There were only three commissioners serving at that time, and they each voted the same way they had in the first place: two voting against the rehearing (thus in favor of the project), and one in favor of a rehearing and opposed to the project.

So FERC was still supporting the project.

When the rehearing was turned down, the same group of project opponents—the attorneys general, the City of Fall River, and the Massachusetts Energy Facilities Siting Board—filed an appeal to the Federal Circuit Court. Filing an amicus brief were the city of Newport and three other Rhode Island towns along the waterway approaching the facility. "The Towns were moved to make this filing because of the unconscionable indifference displayed to their interests, and to the health and well-being of their citizens," according to the brief.

In addition, the Conservation Law Foundation and a local resident, safety engineer Michael Miozza, each made separate appeals to the circuit court. Miozza is concerned not only about safety and potential environmental impacts, but also about the approval process. As they had at Long Beach, FERC denied public access to significant portions of the project's EIS, labeling it "critical energy infrastructure information" and, therefore, of national security significance. Miozza filed under the Freedom of Information Act to see the documents, signing the required nondisclosure agreement. FERC sent the documents to him—by ordinary mail—and so much was blacked out that, essentially, they added nothing to the public record, says Miozza.

On February 2, Weaver's Cove Energy announced a solution to the bridge problem: smaller ships. They would use vessels with a 55,000 cubic meter capacity, rather than the 145,000 cubic meters that they had first proposed. These ships would be 725 feet long and only 82 feet wide, with a loaded draft of 36 feet; they would be able to pass under the old bridge. The vessels would arrive every three days—120 deliveries a year, instead of the previously predicted sixty.

Navigation would remain tricky, though; there is barely 1,100 feet between the old Brightman Street Bridge and its new replacement. The two bridges don't line up, either; a vessel shaped up to pass beneath the old bridge finds herself pointed directly at the abutment of the new one. The company says it's confident that this is not a problem. They arranged to have simulation studies performed that demonstrated that the "unusual maneuver" between the bridges was possible. The simulation also helped them determine the dimensions of the smaller ships. "Navigation through this area is feasible with certain limitations," says the report of their consultants, Marine Safety International (whose simulator we've met before). "After line up and transit of the first bridge and stopping the ship between the bridges, the ship must be moved latterly, by the tugs, the ship's engine and rudder, approximately one beam width until aligned with the opening of the second bridge in the sequence." (It is clear from the diagrams included with the report that they meant "laterally.") Ships do not like to go sideways, particularly long, heavy ships, but these carriers would have to do just that.

After describing the exercise itself, the Marine Safety International report adds the comment, "Maneuvers that require ships to be driven directly at objects while at speed will be inviting trouble." It goes on to discuss tugs (a minimum of three) and, finally, the environmental conditions under which the trick could be performed. Their simulation assumed slack water conditions and a northwest wind of ten to fifteen knots. In higher winds, they say, the passage would be unadvisable.

Slack water comes at long intervals and for short periods of time, which may or may not coincide with the calm winds required for passage between the two bridges. Interestingly, there was no indication that they had simulated unberthing and heading back down through the two bridges. From the diagrams, it would appear that that maneuver, too, would be difficult, and there's very little leeway when approaching the old bridge, with but sixteen feet to spare.

An operator of tractor tugs in another port assures me that with the right equipment, the job could be done. "It would be tricky," he says.

"There's gonna be days where nothing moves." With 120 deliveries planned each year—240 transits—how many lay days can be allowed?

It has been suggested that the company's change in plan to use smaller vessels may simply have been a bargaining tool or other ploy to keep the project alive. The opposition immediately took the stance that doubling or trebling the number of trips changed the proposition entirely, and therefore FERC should mount an entirely new review. Probably most significantly, the Coast Guard, stating that there is no margin for navigational error with the new scheme, asked for a revised plan. A week or so after that Coast Guard comment, the Army Corps of Engineers joined the call for FERC to thoroughly review the new plan. But all that review changed nothing. The plan is still in go-ahead mode.

Meanwhile, the governors of Massachusetts and Rhode Island, senators Kennedy and Kerry, and several other members of the New England congressional delegation made an unsuccessful call for a regional approach to siting LNG facilities. (Nothing came of this; FERC is happy with things as they are.)

In July 2006, Rhode Island enacted a law that would prevent LNG carriers from traveling through parts of the approach to Fall River. Passed unanimously in the house and by a margin of more than two to one in the senate, the new bill was based on the federal Waterfront Safety Act which, among other things, provides for the establishment of safety requirements to protect the shore along navigable waters. Rhode Island's new law specifically prohibits a broad spectrum of waterfront activities and structures, as well as all persons, within the Coast Guard–set safety and security zone extending two miles ahead, one mile behind, and a thousand feet on either side of an LNG tanker. The zone applies on land as well as water, and since the route passes within a thousand feet of shore in several places, the new law would prevent LNG carriers from passing. The new law is, of course, expected to be challenged in court.

FERC commissioners have said on numerous occasions that they do not favor projects that do not have local support. It is hard to see that their actions in Fall River demonstrate that position.

Easier Going on the Gulf Coast

Not all applications for new LNG terminals in the United States face so much opposition. One that did not is the Cameron project near Lake Charles, Louisiana.

The last census says Lake Charles is a city of about 72,000 residents in an area totaling 184,000, but officials believe the population has grown 10 to 15 percent since Hurricane Katrina. It lies thirty-four miles north of the Gulf of Mexico, and is served by a ship channel along which there has been a great deal of industrial development over the years; since World War II, this has centered on refineries and other oil, gas, and petro-chemical endeavors. From Sabine Pipe Line's Henry Hub, in nearby Erath, Louisiana, pipelines connect to the Midwest, Northeast, and Southeast as well as along the Gulf Coast. Henry Hub is the primary pricing point for national spot and futures natural gas prices. The area around the Gulf of Mexico is the heart of the natural gas business east of the Rockies, and people are well-accustomed to industrial operations in their vicinity.

There has been an LNG terminal, Trunkline, in Lake Charles since the 1970s. That plant has built an industry track record locally. It has expanded twice in recent years and has yet another expansion proposed. The new Cameron facility is fifteen miles south of Lake Charles and eighteen miles north of the Gulf Inland Waterway, which passes through Lake Calcasieu.

"LNG terminals seem innocuous there—clean and quiet—and people welcome them," says Brad Floyd, who as part of the consulting company TRC Solutions served as project manager for Cameron as it went through the application phases. The first of the rash of stateside projects for new terminals to come before FERC since 2001, the Cameron project was orig-inally known as Hackberry. The application was made at the end of May 2002 by Sempra, which has other projects under way in Baja California and Port Arthur, Texas.

There were some issues to be considered—people were concerned about safety and liability and the impact on the channel, both as to traf-fic and erosion, but these were all addressed early on. One issue that, in general, continues to cause consternation around the Gulf is the effect of pipeline construction in wetlands, but even this was resolved reasonably smoothly in the case of the Cameron LNG plant application. The final EIS was issued just over a year later, and Brad doesn't remember any changes having to be made between the draft and the final version.

Brad Floyd has been involved in eight or nine LNG terminal applica-tions so far, including another particularly easy one in Port Arthur, Texas, an area that has been depressed of late. Its refineries have provided its backbone since the 1920s, and those aging facilities are not the most clean or efficient. New industry is welcomed in the area, where there are now three LNG projects.

Those Opposed

IN EVERY single place that applications have been considered for building LNG import terminals in the United States, there have been at least a few opponents to the proposals. Some object to one or more particular facilities, and others are against any LNG imports at all. They state a variety of reasons for their positions, some of them entirely reasonable, and some less so.

It's easy to dismiss people who have a blanket opposition to anything. Says Neal Frangesh, "NIMBY—Not In My Backyard—was replaced by BANANA: Build Absolutely Nothing Anywhere Near Anyone. Now we've got NOPE: Not On Planet Earth." There is an incongruity when people espousing any of those positions against energy sources continue to turn on the lights in their houses. Some individuals have even been known to suggest that more electricity should be used, as it causes no pollution. Where do they think those kilowatts come from?

One doesn't mind knowledgeable, thoughtful people taking a stance against something. Many people in California, for instance, believe that by conservation and intelligent, careful use of the energy resources they have, plus serious research and investment in renewables, the importing of more natural gas can be avoided. It may be so in California, and there is certainly an appeal to the argument. No one wants to be dependent on other nations, particularly those with which we may have political disagreements. Some Californians also see disruption in LNG deliveries as being the possible cause of increasing electric rates or decreasing availability of electricity. It's easy to understand why Californians in particular would be concerned about the potential for manipulation of energy prices—once burnt, twice shy. Their victimization by Enron and other unscrupulous commodities-price manipulators in 2000 and 2001 was a huge warning beacon.

For people in New England, however, whose electric prices are 36 percent higher than those of the rest of the country,[96] that argument is less convincing. Not that an increased supply of natural gas will necessarily lower prices, despite what some hope or even suggest, but perhaps its increased availability would at least prevent prices from rising too much more, too quickly.

There are also reasonable objections to LNG itself. While it certainly has an amazingly clean record of safety, there is always the lurking threat of accident—or worse yet, terrorism. There has not been an actual LNG event of the scale that can be hypothesized, so no one fully knows what would happen in such a case. Some see LNG as potentially attractive for terrorists, and the consequences of a successful terrorist attack on a natural gas facility could be of great magnitude. Many residents of urban areas where LNG terminals are proposed are worried about domino effects, one plant setting adjacent facilities or plants afire, releasing chemicals or other petroleum products or causing explosions. No one wants a large fire nearby, no matter where they live, but in a highly populated area, or one of great economic significance, obviously the consequences could be devastating.

Some opponents object to the hassle that LNG traffic causes. The fact that carriers require security areas around them as they approach their berths does inconvenience others on the water and even people in vehicles if traffic must be stopped while LNG tankers pass beneath bridges.

The concern about ecological harm from open-loop vaporization systems is reasonable, too, as are worries about air pollution from the ships' and tugs' exhaust, or threats to whales or other species that may feed or breed in the pathway of the LNG vessels. Also understandable are the questions about narrow waterways and hazardous currents, both from a navigational standpoint and from concern about the effect onshore if a vessel did get in trouble.

In addition, many residents don't like changing the character of the area in which they live. The facilities themselves do have an industrial look to them, which people find offensive in relatively pristine areas. (What many people do not realize, however, is that today's stretch of "unspoiled" coast might well have been a busy commercial or industrial site a few generations back. The area around Downeast LNG's proposed site in Robbinston, Maine, has been home to a shipyard, brickyard, and shingle mill in the past, while other parts of Passamaquoddy Bay have served a number of canneries and, once, a cotton mill.) It's undeniable, though, that an LNG terminal is on a different scale than most rural areas are accustomed to. In tourist areas, some worry that the presence of LNG ships and terminals will discourage visitors. There is fear that one industrial use will inevitably lead to more. Offshore placements also are questioned from an aesthetic point of view as well as from concern about cluttering the ocean with impassable zones.

Then there are the people who fear that property near a facility will lose market value. (This does not seem to have been the case so far in Everett, Massachusetts, which has shared the rapid increase in real estate values that Greater Boston has experienced in recent years. Property on Boston Harbor has not suffered from its proximity to the Distrigas facility or the ships that pass close offshore, judging by the new million-dollar condominium units all along the shore. The idyllic village of Wiscasset, Maine, was home to the Maine Yankee nuclear plant, and yet property values in that town are in line with those of the towns around it.)

Laypeople and researchers alike are concerned about whether FERC requirements and other regulations are adequate to protect the public, and some are cynical about what they see as the government's uneven—and sometimes erroneous—application of those regulations.

Individuals as well as organized groups, large and small, are fighting LNG on all coasts, often making good use of online mail lists and newsgroups. There are many Web sites about LNG. Some are from the LNG industry and depict a utopian vision. Some are from organizations honestly trying to show a realistic picture, with links to all kinds of studies and news stories and to government and industry sites as well as anti-LNG sites. Still other Web sites are rabidly anti-LNG, and some of these don't mind posting outdated information, implying untruths, and throwing in red herrings specifically to frighten people.

Sometimes, the groups fighting a particular project may have a motive beyond what shows up in their speeches and printed material, or even beyond what is felt by their individual members. For example, it's rumored that Canadian groups fighting the Passamaquoddy Bay projects in Maine have financial backing from Irving Oil, which is constructing its own LNG import terminal in St. John, New Brunswick. If a plant is built in Maine, the St. John facility could lose a significant part of its market.

A complex debate can make for strange bedfellows. A meeting held by Quoddy Bay LNG during the winter of 2006 on Campobello Island, New Brunswick, was well attended by fishermen and others concerned about fishing, fishing gear, endangered species, and the safety of bringing large tankers through the very narrow, rock-delimited Head Harbour Passage, where the currents are strong and peculiar. "I never thought I would be on the same side as the Sierra Club," grumbled one fisherman.

Probably the most widely known warrior against LNG in general is attorney Tim Riley, who lives in Oxnard Shores, California. Riley is a civil litigator by trade—his Web site, TimRileyLaw.com, advertises, "If you or someone you care about has sustained a serious personal injury, or has been treated unfairly by an insurance company, then call Attorney Tim Riley for a free telephone conversation." He provides links to specifically anti-LNG Web pages, too, filled with horror stories and sensational and repeated composite images showing square miles of flames superimposed over suburban California landscapes.

Riley describes his first introduction to LNG, a few years ago. "I received some mail addressed to 'Community Leader,' a high-end, quality brochure talking about a proposed LNG facility off our beach. They were boasting that they were going to be able to provide 25 percent of all California's residential natural gas needs, and they showed their facility looking harmonious with the fish, the sea creatures happy to be involved with the complex," he says. "It looked fishy. I decided I'd better look into it."

He says that he went at his research just as he would have done for one of the large personal injury cases he handles—as if he were going before a jury. The more he learned, the less he liked LNG. He came across photographs of the Cleveland disaster, which he'd never heard of before. "No one had. I saw pictures of it, and it looked like Hiroshima after the bomb." That was when Riley realized he had a role to play. "I was like Paul Revere, 'The LNG is coming, the LNG is coming!'" He read the environmental impact report from a 1977 application for an import terminal in Ventura, California, which never came to fruition. There he saw models of possible vapor plumes, extrapolated directly from the small-scale tests that had been performed to date, which predicted flammable vapor cloud travel distances of 30 miles, 50 miles, even 127 miles. He was horrified.

Riley's research stopped there, with that thirty-year-old report, although that of other entities has not. "The laws of physics haven't changed since 1977," his Web site states. When I said to him that no one believes those estimated vapor plume distances anymore, Riley's answer was simple: "I don't really care." He says that for his graphic, he "chose the most conservative one: thirty miles."[97] (The most conservative of the 1977 predictions, that is. The 2004 Sandia study's worst-case scenario predicted that, under the most favorable dispersal conditions, a plume of natural gas could extend perhaps as far as 2.25 miles from its source before becoming too diffuse to ignite.)

Riley's Web site is dramatic, with red headlines that jump out at the viewer scrolling down the long pages. Along with the photographs of the devastation of Cleveland, Riley has some pertinent, informative articles and links to good and impartial resources such as the Sandia report. Many of his sources are credible—the *Los Angeles Times*, the *Boston Globe*, even *Oil & Gas Journal*. But there are some other pieces, too, including Rhode Island representative Patrick Kennedy's hypothetical exploding LNG facility that "would decimate a 50-mile radius." That phrase is highlighted in heavy red type, as is the headline about the "LNG pipeline" that exploded in Nigeria. Riley justifies these by saying that he's simply quoting news stories (or the Congressional Record), which is true. In the Nigerian case, farther down the page in small black letters, the article says that the line in question ran from the gas field into the liquefaction plant, so although it obviously was not carrying liquefied gas, Riley can't quite be accused of not telling the whole story. Not quite.

Riley and his wife say they are making their fight pro bono and have taken no donations for the project. "It's out of our own good hearts, against the richest industry on Planet Earth. It's something our fellow Americans have to know about." They have produced a movie innocuously titled *The Risks and Dangers of LNG,* which they promote throughout the Web site and which they say provides them with the only income they make from their crusade. It sometimes finds its way to communities who think they're getting an unbiased view of LNG. That is not the case.

Like many opponents of LNG, and there are thousands of them, the Rileys are not all wrong, nor are they all right. They perhaps do a disservice to the quieter groups who sincerely want to do the right thing. Some of those have been successful locally in chasing the big oil companies away, as they have done in both the Northwest and the Northeast. No one yet has stopped the whole show.

Bottom Line

THE United States is facing an energy gap—a gap as petroleum and domestic natural gas become scarcer while the next fuel has yet to become viable on a wide-scale basis. One hopes that the new fuel, whatever it may be, is a renewable and clean resource.

Some think that hydrogen can fill the bill, while others say it hasn't much chance, depending as it does on a second energy source, which to some degree puts us back in the same predicament we're in today. In any case, the technology for hydrogen isn't developed to the point of universal usefulness yet. The only other large-scale option open to us any time soon is probably nuclear power, which may see a resurgence in the United States, as it has in some other countries. (At the same, time it's being pulled back in still other places, such as Sweden.) Barring failures along the line of Chernobyl or even Three Mile Island, nuclear power has major ecological advantages. There is the matter of what to do with spent fuel rods, which (though we've been doing so for years) is hard to ignore. But even nuclear power—if, indeed, it's a good solution—isn't going to fill the energy gap immediately. Public perceptions cannot be changed with such alacrity. Anyone who thinks the permitting process for an LNG facility is difficult should look at what is involved in getting a nuclear power plant approved!

Some say that it makes no sense to push for what can only be a stop-gap measure, but something must be done, and increasing importation of LNG onto our shores is certainly a logical option. It's not as if the energy gap would be of such short duration. Many LNG plants are being designed for a forty-year life span, and that seems an entirely reasonable span to cross the gap.

We're already using a lot of natural gas, with apparent aplomb. Importing more is simply adding to our supply of that commodity. The only increased peril from liquefying the stuff comes in handling it when it's concentrated or in transition. Liquefaction, transportation, and regasifying all have potential hazards. The liquefaction step isn't our problem, for the most part. We have the one export facility in Alaska, but that's been doing its thing uneventfully for several decades now, and there's no reason to think it won't just keep on as it has been. It has no likelihood of making any impact on the energy situation in the lower forty-eight, anyway. Small liquefaction plants exist today in various parts of the country, but they're not increasing our energy supply, merely storing it. Regasifying has been going on with little difficulty at each of the peak-shavers and in four import facilities in operation around the country for awhile now—more than three decades without significant interruption, in the case of the Everett, Massachusetts, facility. The technology of liquefaction and regasification is well understood. Transportation of LNG (and other liquid gases, for that matter) also has an excellent safety record, and it seems reasonable to expect that, barring malicious interference, this will continue, though the looming shortage of qualified merchant marine officers raises some questions.

What's an Acceptable Risk?

The answer certainly isn't entirely black or white when attempting to determine what is an acceptable level of risk relative to the hoped-for level of benefit. Given the desirability of a particular end—in this case, that there be more energy available—does the importing of LNG provide a solution in a manner that is as acceptable, or more so, than some other manner? What are the benefits of LNG compared to, say, coal or nuclear energy? What are the dangers, compared to other schemes? And what are the probabilities of situations with dangerous consequences actually occurring? What actions are being taken to minimize the probabilities of such events and mitigate the dangers if the events should, in fact, take place?

Nearly everything we do has some risk attached to it. I used to live about a half-mile down the road from an elderly man whose mailbox was set across the road from his house. Six days a week for many years, he crossed the road without incident. Then, one evening, a car struck him, and he died. Should he not have ever crossed that road, because he might get killed? Of course he was risking only his own life, although I suspect the life of the young woman who struck him won't be quite the same ever again, either, even though she was found to be legally without fault. But she did make the choice to drive her car, accepting whatever risks went with that. Members of the public who are placed at increased risk—however small—by a new LNG plant will not have made a choice in the matter.

The newspapers report commercial airline accidents regularly, accidents that kill hundreds of people at a time. Occasionally, and sometimes in the most lethal cases, the crashes are caused by terrorism, but far more common are incidents caused by pilot error or mechanical failure. Worldwide, as of October 2005, there had been more than a hundred crashes that had killed more than 125 people; the earliest was a collision over Staten Island in 1960, when 128 died on board, along with six on the ground. Certainly the most devastating of all plane disasters took place on September 11, 2001, but hijacking, sabotage, and terrorism only account for 5 percent of all fatal crashes from the 1950s through 2004. Before September 11, 2001, we would have assumed that very nearly all the risk was taken by the travelers, but that day clearly demonstrated that many people on the ground can be put in danger by an errant airplane. Still, the odds of being killed in an automobile are about two and a half times greater than in a commercial airliner,[98] and everyone drives everywhere. We all take known risks all the time.

Inevitably, if one's talking with mariners about any dangers associated with LNG carriers, someone will mention other vessels that they see as being more nearly like bombs awaiting detonation—carriers of jet fuel and kerosene, for instance, or liquefied petroleum gases, which usually travel under pressure. Unlike LNG, some of these products are, in fact, explosive in unconfined spaces, and they're everywhere.

Chlorine, anhydrous ammonia, unrefined benzene products, naphtha, and phenol are interesting and frightening cargoes, too—the atmospheric impact that any of these chemicals, let loose, could have on their surroundings is unquestionably horrendous to contemplate. These all travel not only by sea but by rail, too, regularly passing by people's backyards all across the country, sometimes stopping on out-of-the-way sidings as other trains go by, and passing through what are seen as dangerous neighborhoods. They are also regularly carried by tractor-trailer on both superhighways and local roads.

Aren't all these barges, ships, railroad cars, and trailer trucks potential terrorist targets? Someone must have deemed them an acceptable risk, rightly or wrongly, and, in fact, they are subject to considerable regulation. But if anyone's calling for movement of these chemicals to stop, the newspapers haven't picked up that story.

Transporting petroleum by water is commonplace in the United States. Not long after the turn of the twentieth century, the first vessel carrying oil from Texas's Sabine Pass headed out with about 3,000 barrels of crude oil. Within a year, more than half the ships coming into the Sabine Pass and Port Arthur were connected with the oil industry.[99] In 2004, well over 260 million barrels of petroleum products of all kinds traveled via tanker and barge from the Gulf Coast to the East Coast, in addition to another million-plus barrels of crude oil. (A further 925,836,000 barrels of refined petroleum products and more than 1,250,000 barrels of crude went by pipeline from the Gulf to the East Coast during that same year.)[100]

And as everyone knows, there have been petroleum spills both from tankers and from pipelines. Since 1970, every year has seen at least five and sometimes as many as thirty-four spills of more than 700 metric tons, as well as between six and ninety-five lesser ones. The numbers are getting smaller as the years go by, averaging fewer than five big ones in the decade starting in 1995, but the accidents still happen.[101]

Standards are becoming stricter all the time. Following the *Exxon Valdez* incident, the U.S. Congress unanimously passed the Oil Pollution Act of 1990, whose intent was to prevent spills, provide a response regimen to clean them up if they occur, and to hold polluters accountable and liable for damages they cause. There are many provisions of the act, but perhaps the single most dramatic one is that tank vessels operating in U.S. waters will all be fitted with double hulls. No new single-hull vessels could be brought into U.S. waters, and existing single-hull tank ships and barges had to be either retrofitted or phased out, the last to be gone by 2015. The double skin gives protection from either the outside getting in, in the case of a grounding that breaches the outer hull (and thus the tank, in a single-skinned vessel), or the inside getting out, if a tank were to fail from within. While undoubtedly there was some grumbling among tanker owners about this and other provisions when the act was passed, the industry has accepted the rules, and everyone in it is more aware of the issues and consequences of pollution. Anyone who even fills a gasoline tank for an outboard motor at a dockside pump has seen the increased precautions, with absorbent towels at hand to sop up the inevitable little drips from the nozzle and a closed container in which to toss the used ones.

It's not all that unusual to hear of explosions aboard petroleum tankers and barges, and of chemical spills from ships and railroad cars. In February 2003, a fuel barge exploded while off-loading four million gallons of gasoline at Port Mobil, Staten Island. Two men were killed. After a railcar overturned in Tennessee in 2002, thousands of gallons of sulfuric acid spilled, and 8,000 people were evacuated from the area. Fortunately, there were no injuries reported. In early 2005, a train wreck in South Carolina caused a chlorine spillage that killed ten. Two hundred fifty more were treated for chlorine exposure, and the 5,400 residents within a mile of the accident were evacuated for nearly two weeks.

While no one wants these things to happen, they do. The risks are deemed acceptable because our modern-day economy and way of life depend on these hazardous products being carried and utilized all around the country.

This is not to say that adding even more potential hazards to our lives is necessarily also acceptable, but rather to point out that we already accept a degree of peril.

I believe that the reason LNG terminals are the cause of so much genuine fear is that—in this country, at least—they are the new kids on the block. (That hardly anyone is even aware of how much LNG already travels about the country in trailer trucks—with a very pure record of safety to date and no one complaining—is almost amusing.) While it has been shipped regularly into the United States since 1971, for most of that time, LNG only came into a couple of ports, so people aren't used to those funny-looking ships that float high on the water. They have heard all kinds of tales and horror stories, some of them true or potentially so, and there are still legitimate questions that have not been answered. Some of those questions can only be answered definitively the hard way—and one hopes they never will be.

There are any number of things we take as given that would never be permitted now, for various reasons. The idea of setting tall wooden poles every few hundred feet alongside nearly every road in the country and stringing wires from one to the next, mile after mile after mile? Crazy! Our electric transmission system is certainly ugly, but even more importantly, the unprotected wires are extremely dangerous when a tree brings one down in a strong wind or heavy snowfall, or when someone touches one with something electrically conductive—a backhoe arm, an aluminum ladder, a sailboat mast, or even a dump truck. Historically, twenty percent of all construction fatalities are caused by electrocution, nearly all from contact with overhead power lines.[102] What's the likelihood of creating such an electricity distribution system anew today? Close to zero, one suspects, but it's taken for granted by us all because it's been like this for as long as anyone now alive can remember.

People write long and tedious essays about acceptable risk, offering one or another way of judging the matter—and there's even a spoof Web site that purports to be corporate America's tool for deciding how much loss of life would be acceptable for how many dollars made. In fact, risk analysis is not a simple matter, and has to consider the benefits as well as the dangers and their consequences, along with their likelihood of happening. It's clear that in the case of LNG import facilities and ships, the potential dangers are both present and possibly horrendous. Depending on the location of a disaster, vast numbers of people and their property could be annihilated. But, certainly speaking from the historical base, these situations have a very low likelihood of arising—considerably less than many other dangerous conditions we've accepted without much concern.

Who is looking out for us in the process of making such decisions?

No one is. This, to me, is the most unfortunate part of the whole situation. There's a long, drawn-out, and complicated government-mandated process to gain a permit for an LNG import facility, with all sorts of documentation required for all kinds of issues, but despite a sequence of federal policies, most recently in the Federal Energy Policy Act of 2005, there is no overall guidance as to where an appropriate site might or might not be. No agency is empowered to do any proactive planning—looking at the country as a whole and saying, "The United States needs to add this many LNG ports, and these are the best locations for them, so don't even bother talking to us about siting an LNG facility anywhere else." This is the most glaring error, among several.

It's not the Coast Guard's job to decide that a particular waterway is not suited to LNG traffic, thereby prohibiting outright the construction of a particular project. "The Coast Guard identifies the resources needed to integrate a proposed LNG facility into the port security plan. FERC has the responsibility for balancing the need for energy with the resources needed to mitigate risks and manage consequences," says Captain Mary Landry.

It's not FERC's job to decide which facilities ultimately get built. "FERC is obligated to conduct an environmental and technological review of all proposed projects," said FERC's Rob Cupina at a 2005 workshop, "but the market and stakeholders will ultimately determine which terminals get built."[103] Despite the statement by former FERC commissioner Nora Brownell that, when the local community is against a project, "there are ways they can stop it," questioners have rarely been given serious credence in FERC decisions. The commissioners themselves, after an initially positive recommendation from staff, determined the turndown in Providence, Rhode Island. Comments made in response to the DEIS forced a review of some calculations, and both staff and the commissioners recognized that the small piece of land on which KeySpan proposed placing its terminal was not adequate for the required exclusion zones. They also declared the grandfathered tank inadequately protective of the public. But unlike the local residents, legislators, attorneys general, and governors, the FERC commissioners were apparently unconcerned about LNG carriers making the twenty-nine-mile transit through densely populated and vessel-filled Narragansett Bay to reach the facility. Nor did they feel that the Weaver's Cove plan for a Fall River facility had a problem with its exclusion zones.[104] Across the country, Long Beach, where a number of state agencies were opposed to the project, was a good test of the power of stakeholders. The experience at Fall River seems to have clearly demonstrated FERC's complete disregard for local concerns, even as expressed by top officials from more than one state.

FERC has seen its role as being simply to ensure that, once built, a project will be as safe as it can be. This approach created a land-rush mentality among potential LNG developers. There has been a frantic push to get applications in, as developers felt (or possibly knew) that the first to come would be the first to be served. There's a demand for more natural gas, but neither the supply nor the demand is unlimited. Many applications were for the Gulf of Mexico, at least in part because the oil and gas business is so well established there that the process is simpler, with fewer objections raised. But is it wise to have so many eggs in that one basket? The hurricanes of 2005 have been a reminder that perhaps that's not a good idea after all.

One of the questions that FERC considers in its environmental impact statement is, what alternatives are there to this particular project before us that can fulfill the same objectives? So far they've shown a pattern of tautologically answering in terms of the project itself, with each EIS concluding that the particular project under discussion is best. An application says its goal is to provide natural gas directly to a particular market, so if another alternative wouldn't provide natural gas directly to that market, but rather by a pipeline, FERC says it wasn't comparable.

Why not actually consider real alternatives? Natural gas can be and is piped from one place to another. An import terminal need not be directly at the market area, and there are often good reasons not to have it there. More than one person has said that if they were given the choice between a facility in Long Beach, California, with a population of more than 461,000, and one twelve to twenty miles off Ventura County's shoreline, they'd certainly opt for the offshore facility—not that there isn't opposition to that siting, too.

Today, many people feel that while increased importation of LNG is a reasonable part of the energy picture for the next several decades, the necessary new import terminals should be placed far from heavily populated areas or offshore. That longtime federal policy was articulated yet again in the Energy Policy Act of 2005. If FERC were to recognize the fundamental importance of this concept and apply it in a commonsense way, then when a project proposed to be sited in an urban area is first presented to them, they could simply say, "No, that location doesn't fit within the guidelines." That, in addition to decreasing the potential impact of a mishap or attack, would save everyone tremendous amounts of money and time—not just the developer, and therefore, the ultimate consumers, but the taxpayers who are supporting all the oversight and analysis and paper generation from FERC and other agencies. As it stands now, FERC purposely makes no general judgments about siting.

It has been said that, until just recently, to apply for a permit was (eventually) to gain one, but that's not quite correct. Either because of the difficulty of the application process, or because potential developers eventually concluded that their idea was not feasible after all, some proposed projects have disappeared along the way. Still, as of this writing, only the one in Providence has actually been turned down by FERC.

LNG ships have transited in and out of Boston Harbor uneventfully for decades, and many, many times more ships have stopped and unloaded in Tokyo Bay. The safety record of LNG being what it is, it's easy to assume that it will continue like that. Certainly, it will be everyone's most sincere intent to maintain that record, it being not only the right thing to do, but also the most economical. Humans will make errors—it's our nature—but the engineers and designers and regulators are making serious efforts to ensure that automated and redundant systems will protect vessels and facilities and their surrounds from damage in the case of error or equipment failure. The March 2006 event at the Elba receiving terminal in Georgia, where an unloading operation was interrupted with nary a drop spilled, shows just how well-engineered these projects really are.

Nothing is foolproof, however, and perhaps most important, post-9/11, we can no longer assume that the worst problems we might face are accidental. We have to focus on malicious intent, and although Incident Action Plans have been created whose authors assure us would be effective, a successful attack on an LNG ship or terminal would make a nasty mess, whether more than or merely as much as some other target. In an urban area, tens of thousands or even hundreds of thousands of people might die. Billions of dollars' worth of property might be lost, and the economy of the nation could be seriously disrupted. That would be its appeal to a terrorist, in fact. The bottom line is, a successful attack in a busy city location would be serious enough that it must not happen.

To this particular student of the issues, the answers to the basic questions have come clear. LNG does fill an energy gap in a reasonable manner, and there are parts of the country that can benefit from it. It's a relatively clean-burning fuel, and it's available. The people who handle it have learned how to do so in a safe manner. Still, like most other forms of energy, it offers significant dangers, even if of extremely low probability, particularly in a day of terrorism. In the words of Harvard physics research professor Richard Wilson, "We must now pay close attention to these events of low calculated probability but high consequence. The terrorists will."[105]

It seems like a no-brainer that LNG ships and facilities should be kept away from population centers and from industrial concentrations that could both attract and magnify the consequences from an LNG incident. Why not locate somewhere that would be far less devastating should something go wrong, and which would be less attractive to terrorists for that very reason? The sorts of locations that make sense are offshore or in rural areas.

Dr. Jerry Havens sums it all up very simply, and, to my mind, entirely correctly: "We want gas, we need gas, and we should get it. But if we have alternatives, we ought not to put it in the middle of town."

Acknowledgments

I have heard complaints that the people arguing that LNG can be managed safely are the same people, again and again. A similar criticism can be lodged about those fighting against new plants. The fact is, there are not very many people with expertise or interest in the field, and I've been fortunate to speak with many of them. My appreciation for those with whom I spoke is immeasurable. Some of them have the educational background or experience to speak authoritatively and some don't. There are but a few people, pro or con, whose motives I question.

I must thank everyone who spoke with me. Some prefer to remain nameless for various reasons, but have been no less important to my understanding of this complicated subject. The others are named in the preceding pages. A few individuals have taken me under their wing, tolerating my naiveté for the most part and bursting idealistic bubbles when they needed to. I am immensely grateful to those people in particular. They know who they are.

Perhaps as much as anyone, the several Coast Guard officers and civilians working with them who gave so freely of their time and information deserve thanks. The Coast Guard is the one group that truly has no interest one way or the other whether more LNG terminals are constructed; it simply wants to make American shipping safe, whatever the commodity or location.

I am saddened to find that so much of the factual material available about natural gas is dismissed so lightly by the foes of LNG as being simply propaganda from the same old sources. There are, of course, legitimate issues to be raised, and I have done my best to raise them, but there are also facts demonstrated in unbiased tests and experiments during the last forty years that many foes will never let themselves accept. Admittedly, the professionals in the field believe in the possibilities for the projects they're planning or designing or operating (or even regulating)—if they didn't, they'd be in some other line of work—but many of them also are willing to listen to and work with people to work out a satisfactory result for all.

I've never had occasion before to spend much time talking shop with engineers. I have to say I've enjoyed just about every minute of the many long conversations I've shared with the engineers in the LNG field. This is a great group of people. In my rummaging around I occasionally found information that was new to a particular researcher or consultant; when this happened, my information was given respect. I admire this open-mindedness on the part of people who already know more than anyone else.

And thanks, Phyllis, for an objective view when I needed it. Hugh, too. And, as always, I appreciate the time my editor, Karin Womer, spent asking pertinent questions and smoothing out my words. This book is better for her involvement.

Notes

Taking a Dispassionate Look at LNG

1. *Newsday*, January 26, 2006, no longer available online.

U.S. Energy Basics

2. All data about energy production, importation, and use not otherwise attributed are from the Web sites of the Energy Information Administration, United States Department of Energy. Estimates of future demand and sources are from their *2005 Annual Energy Outlook*.

3. PBS broadcast, "The First Measured Century," hosted by Ben Wattenberg, http://www.pbs.org/fmc/book/14business8.htm#buschart8/.

4. "Crude Oil Prices 1861–2005," Forbes.com, http://www.forbes.com/static _html/oil/2004/oil.shtml/.

5. John H. Wood et al., "Long-Term World Oil Supply Scenarios," Energy Information Agency, Department of Energy, August 18, 2004, http://www.eia.doe .gov/pub/oil_gas/petroleum/feature_articles/2004/worldoilsupply/oilsupply04.html/.

6. Sara Banaszak, "North American Resources and LNG Imports Can Meet Future Demand," presentation at National Petroleum Council, 2003.

7. Environmental Protection Agency Web site, http://www.epa.gov/history /topics/perspect/london.htm; also Department of Energy Web site.

8. Statistics from the World Nuclear Association, http://www.world-nuclear .org/info/reactors.htm/.

9. UC Berkeley, "Ethanol can replace gasoline with significant energy savings, comparable impact on greenhouse gases," press release, January 26, 2006.

10. Jeff Johnson, "Power from the Sun," *Chemical & Engineering News,* June 21, 2004, http://www.pubs.acs.org/cen/coverstory/8225/8225solarenergy.html/.

11. "What FERC Does," http://www.ferc.gov/about/ferc-does.asp/.

Natural Gas from the Beginning

12. Account primarily from John R. Hale et al., "Questioning the Delphic Oracle," *Scientific American*, August 2003, http://www.sciam.com/article.cfm?articleID =0009BD34-398C-1F0A-97AE80A84189EEDF/.

13. Byron W. King, "The Ghost of Colonel Drake," *Energy Bulletin*, November 7, 2004, http://www.energybulletin.net/8295.html/.

14. Daniel Yergin, *The Prize*, New York: Simon & Schuster, Touchstone, 1992, p. 28.

15. Ibid., p. 30.

16. Account primarily from Jim Bradshaw, "Peaceful Lake Peigneur Turned Into Maelstrom," *Lafayette* (LA) *Daily Advertiser*, November 25, 1977.

17. J. Stephen Morrisson, of the Center for Strategic and International Studies, speaking to a Senate committee in 2004, according to an NPR report by Steve Inskeep, August 25, 2005, http://www.npr.org/templates/story/story.php ?storyId=4797953/.

Aspects of the LNG Distribution Business in the United States

18. "Annual Energy Outlook 2006," EIA, U.S. Department of Energy, http://www.eia.doe.gov/oiaf/aeo/pdf/aeotab_13.pdf/.

19. Bureau of Transportation Statistics Web site, http://www.bts.gov /publications/national_transportation_statistics/2004/html/table_01_10.html/.

20. American Gas Foundation, "Safety Performance and Integrity of the Natural Gas Distribution Infrastructure," March 2005, www.gasfoundation.org /ResearchStudies/safety_perf.htm/.

21. W. Jacqueline Kious and Robert I. Tilling, "This Dynamic Earth: The Story of Plate Tectonics," 1996, http://pubs.usgs.gov/publications/text/dynamic.html/.

22. Gordon Weil, "Kenai Peninsula Report," October 30, 2003, http://www .harpswell.info/frwds/files/kenai-report.htm; Jed Watkins, interview with author, October 2005.

23. California Electricity Oversight Board, "The Current Roles of Natural Gas and LNG in California," 2004, http://www.eob.ca.gov/attachments/081004NatGas Report.pdf/.

24. Razi Syed, "Automobile Special: Pakistan Is Third-Largest User of CNG," *Daily Times*, Pakistan, November 27, 2005, http://www.dailytimes.com.pk/default .asp?page=2005%5C11%5C27%5Cstory_27-11-2005_pg5_3/.

25. Mike Hubbell, interview with author, November 16, 2005.

Notes

LNG Carriers and Their Crews

26. "World LNG Shipping Capacity Expanding," EIA, U.S. Department of Energy, 2004, http://www.eia.doe.gov/oiaf/analysispaper/global/worldlng.html/.

27. Maritime Business Strategies, LLC, "The Orderbook of LNG Carriers," December 24, 2005, http://www.coltoncompany.com/.

28. "Simply the Best," Moss Maritime promotional brochure.

29. Hisashi Yamamoto, "Manning the Ship," *Coast Guard Proceedings*, USCG, Fall 2005, pp. 47 ff.

30. Ben Raines, "Huge Gulf Oil Spill Blamed on Rita," *The Times-Picayune*, New Orleans, January 21, 2006; personal correspondence with Capt. William Sullivan, Operations Manager, K-Sea Transportation.

31. "Judgments of Four Major Marine Accidents Inquiries during the Past Half-Century in Japan,"http://www.nippon.zaidan.info/seikabutsu/1999/00813/contents
/020.htm/.

32. Other information for this report from "Focus on IMO," International Marine Organisation, May, 1996, http://www.imo.org/includes/blastDataOnly.asp /data_id%3D7999/IMDGdangerousgoodsfocus1997.pdf; Cornwell, John, "Modeling LNG Spills in Boston Harbor," 2004, http://www.ocean.oce.uri.edu/oce495/Quest_LNG %20spill%20report%202003.pdf/.

LNG Characteristics and Hazards

33. Center for Liquefied Natural Gas, "LNG Carrier Safety," http://www .lngfacts.org/About-LNG/Carrier-Safety.asp.

34. Dr. Jerry Havens and Tom Spicer, "LNG Vapor Cloud Exclusion Zones for Spills into Impoundments," *Process Safety Progress,* American Institute of Chemical Engineers, September 2005.

35. Mike Hightower et al., "Guidance on Risk Analysis and Safety Implications of a Large Liquefied Natural Gas (LNG) Spill over Water," Sandia National Laboratories, December 2004 (henceforth referred to as "Sandia"), p. 51.

36. Sandia, p. 53.

37. Report on Montoir Tests (France), 1989, in Sandia report, pp. 123–24.

38. Dr. Alan Schneider et al., *U.S. Coast Guard Liquefied Natural Gas Research at China Lake,* U.S. Department of Transportation, 1980.

39. Dr. Jerry Havens, "LNG and Public Safety Issues," *Coast Guard Proceedings,* Fall 2005, p. 31.

40. Sandia, p. 39.

41. Ibid.

LNG-Related Accidents (and Those Blamed on LNG)

42. Saïd Aït-Hatrit, "Catastrophe of Skikda," news report, Afrik.com, January 20, 2004, (translation by Google), http://www.afrik.com/article6961.html/.

43. California Energy Commission staff, "Algerian LNG Plant Explosion Fact Sheet," April 20, 2004, http://www.energy.ca.gov/lng/news_items/2004-01_algeria_factsheet.html/.

44. Bachir Achour and Ali Hached, "The Incident at the Skikda Plant: Description and Preliminary Conclusions," *Sonatrach*, March 2004, http://www.energy.ca.gov/lng/news_items/2004-03-21_SONATRACH.PDF/.

45. California Energy Commission report.

46. Nathan Siu et al., "Qualitative Risk Assessment for an LNG Refueling Station and Review of Relevant Safety Issues," Idaho National Engineering Laboratory, Alternate Fuels Program, Lockheed Martin Idaho Technologies Company, Idaho Falls, Idaho, prepared for the U.S. Department of Energy, 1998, pp. 76–77; New York State Energy Planning Board, "Report on Issues Regarding the Existing New York Liquefied Natural Gas Moratorium," November 1998, pp. 6-2, 6-3.

47. CH-IV International, "Safety History of International LNG Operations," January 2005, p. 8.

48. Shell Oil's Web site: http://www.shell-usgp.com/lngsashas.asp; New York State Energy Planning Board report.

49. Nathan Siu et al., *Qualitative Risk Assessment*; Lee Niedringhaus Davis, *Frozen Fire*, Friends of the Earth, San Francisco, 1979, pp. 98–104.

50. Dr. Jerry Havens, interview with author, March 8, 2006. He was sent by the U.S. Coast Guard to Algeria immediately after the accident to research its causes.

51. Much of the information about the various incidents, including this quotation, is from CH-IV International, "Safety History."

52. CH-IV International, "Safety History."

53. http://www.TimRileyLaw.com, "LNG Danger to Our Communities."

54. Tim Riley, interview with the author, February 16, 2006.

55. "Safety Reports, Resources & Documentation about LNG & Pasamaquoddy Bay," http://www.quoddyloop.com/savepassamaquoddybay/safety_reports.html/.

56. Media release from the International Federation of Rock Art Organizations, http://www.chinadaily.com.cn/en/doc/2003-12/26/content_293435.htm/.

57. CNN.com, "Thousands flee China gas disaster," December 26, 2003, http://www.cnn.com/2003/WORLD/asiapcf/east/12/25/china.gas/.

Threats

58. Much of this chapter is based on figures from the EIA, U.S. Department of Energy, and an interview with Aloulou Fawzi at that organization, March 1, 2006.

59. H.E. Abdullah Bin Hamad Al-Attiyah, Second Deputy Premier, Minister of

Notes

Energy and Industry, Qatar, "The Role of Qatar in the Global LNG Market," speech at the Energy Policy Foundation Conference, Sanderstolen, Norway, February 4, 2005, http://www.mobile.qp.com.qa/qp.nsf/53e8868bdf4e6bf3432569c2003b31e9/b96b2810 c9045acd43256fa3004b9e54?OpenDocument/.

Considerations of Terrorism

60. From U.S. Congressional Record, Arguments over Energy Bill, H.R.6 Energy Policy Act pf 2005.

61. Lord Levene, speech to Houston Forum, September 20, 2004, quoting Dr. J.C.K. Daly, online source no longer available.

62. Lloyd's Register, "Statement on LNG Risks from Lloyd's Register North America, Inc.," September 23, 2004, http://www.lr.org/news/press_releases/2004/pr_0924_lng_statement.htm/.

63. Dr. Phani Raj, "LNG Safety: Myths and Facts," presentation at NECA Conference on Natural Gas Issues, September 20, 2005, http://www.necanews.org/index.php?option=com_docman&task=doc_download&gid=269-/.

64. Richard A. Clarke et al., "LNG Facilities in Urban Areas," Good Harbor Consulting, LLC, May 2005, p. 3.

65. Ibid., p. 4.

66. International Maritime Organisation press release, "IMO to Take Straits Initiative," November 11, 2004, http://www.imo.org/Newsroom/mainframe.asp?topic _id=848&doc_id=4466/.

67. Ioannis Gatsiounis, "Malacca Strait: Target for Terror," *Asia Times* online, August 11, 2004, http://www.atimes.com/atimes/Southeast_Asia/FH11Ae02.html; "Pirates Mock Malacca Strait Security," *Asia Times* online, April 9, 2005, http://www.atimes.com/atimes/Southeast_Asia/GD09Ae02.html/.

68. Much of my overall discussion of piracy comes from John S. Burnett, *Dangerous Waters: Modern Piracy and Terror on the High Seas*, Dutton Books, 2002.

69. Charles Glass, "The New Piracy," *London Review of Books*, December 18, 2003. http://www.lrb.co.uk/v25/n24/glas01_.html/.

70. Captain Dick Johnston, interview with author, March 11, 2006; corroborated by Robert Lakey, interview with author, March 24, 2006.

71. Dr. G.A. Melhem et al., "Managing LNG Risks: Separating the Facts from the Myths," Iomosaic Corporation, 2005.

72. GAO Report to Congressional Requesters, "Homeland Security: Voluntary Initiatives are Under Way at Chemical Facilities, but the Extent of Security Preparedness is Unknown," March, 2003, p. 1.

73. Ibid., p. 10.

74. John Nordin, PhD, "Initial Isolation and Protective Action Distances, Plume Modeling," *The First Responder*, January, 2005, http://www.imakenews.com/aristatek/e_article000348237.cfm?x=b4qpwVc,b2jjkn3J,w/.

75. The Bhopal Medical Appeal and Sambhavana Trust, "What Happened in Bhopal?" 2004, http://www.bhopal.org/whathappened.html/.

76. CBC Web site, "The Halifax Explosion," 2003, http://www.cbc.ca /halifaxexplosion/; David A. Greenberg, "Modelling the Tsunami from the 1917 Halifax Harbour Explosion," *Science of Tsunami Hazards*, 1993.

77. The Ocean Channel, "Transporting Hazardous Chemicals by Sea," 2003 data, http://www.ocean.com/resource.asp?resourceid=4008&catid=5&locationid=2/.

Listening to Stakeholders: The Licensing Process

78. Alan Moore, e-mail message to author, January 20, 2006.

79. U.S. Department of Homeland Security, "Protecting America's Ports, Maritime Transportation Security Act of 2002," press release, July 1, 2003.

Standards

80. Energy Policy Act of 2005, Title III Subtitle B, Section 311 (d).

81. Rob Cupina, interview with author, March 10, 2006.

82. M. Sam Mannan et al., "LNG Safety—An Update on Recent Issues," Business Briefing, *LNG Review,* 2005, p. 51.

83. Dr. James A. Fay, "Public Safety Issues at the Proposed Harpswell LNG Terminal," 2003, http://www.harpswell.info/frwds/files/fayreport.htm/.

84. M. Sam Mannan et al., "LNG Safety," p. 49.

Two Contrasting Projects

85. "About SES," http://www.soundenergysolutions.com.

86. "About ConocoPhillips: Who We Are," http://www.conocophillips.com /about/Who+We+Are/index.htm/.

87. Unless otherwise credited, information about the Port of Long Beach is from their Web site: http://www.polb.com/default.asp/.

88. California Public Utilities Commission.

89. California Energy Commission, 2005 Integrated Energy Policy Report, http://www.energy.ca.gov/2005_energypolicy/index.html/.

90. California Energy Commission, "Safety Advisory Report on the Proposed Sound Energy Systems Liquefied Natural Gas Terminal at the Port of Long Beach," September 2005, p. 10.

91. FERC, "FERC Affirms Its Approval of Weaver's Cove LNG; Stands by Its Rejection of KeySpan LNG in Providence," press release, January 19, 2006.

Different Regions, Different Results

92. MassStats, Interactive Maps and Statistical Views of Massachusetts, http://www.merlot.caliper.com/Maptitude/MassStatsNETVersion/(2zhmo0mx2udt2r45pvltv445)/Map.aspx/.

93 Amicus brief of Rhode Island cities and towns filed with FERC, September 26, 2005, on record in the Weaver's Cove file at FERC.

94. Letter from Mayor Edward M. Lambert, Jr. to Mr. Patrick Wood, Chairman, FERC, Weaver's Cove file at FERC.

95. Letter from Captain Roy A. Nash, USCG, to Captain Michael W. Byman, USN, filed as part of the Naval Undersea Warfare Center Division, "Withdrawal of Motion to Intervene, Motion to Reopen, and Request for Rehearing in the Application of Weaver's Cove Energy, LLC," Weaver's Cove file at FERC.

Those Opposed

96. EIA, U.S. Department of Energy, January 2006.

97. Tim Riley, interview with author, February 16, 2006.

Bottom Line

98. Office of Hazardous Materials Safety, "A Comparison of Risk, Accidental deaths—United States—1994–1998," U.S. Department of Transportation, http://www.hazmat.dot.gov/riskmgmt/riskcompare.htm/.

99. Historic marker, Port Arthur, Texas, erected 1969 by the Moody Foundation, http://www.co.jefferson.tx.us/Historical_Commission/markers/Tanker_Service_Early_Oil.pdf/.

100. U.S. Energy Information Administration, Department of Energy.

101. International Tanker Owners Pollution Federation Limited, http://www.itopf.com/stats.html/.

102. "Electrocution," *Construction Safety*, Spring 2000, http://www.csao.org/uploadfiles/magazine/vol11no1/shock.htm/.

103. Pipeline and Hazardous Materials Safety Administration, "Summary, LNG Community Awareness Workshop, Washington DC, February 2, 2005," http://www.primis.rspa.dot.gov/meetings/mtg29/LNGCommunityAwarenessWorkshopSummary-Feb2-2005.pdf/.

104. FERC, "Commission Approves Proposed LNG Terminals in Texas and Massachusetts; Rhode Island Terminal Rejected," press release, June 30, 2005, http://www.ferc.gov/press-room/press-releases/2005/2005-2/06-30-05-lng.asp/.

105. Richard Wilson, "Combating Terrorism: An Event Tree Approach," talk given at the University of New Mexico (Los Alamos), 2003.

Index

Index

231

Index